# On the Wing

# On the Wing
*Insects, Pterosaurs, Birds, Bats and the Evolution of Animal Flight*

David E. Alexander

Illustrations by Sara L. Taliaferro

Oxford University Press is a department of the University of Oxford. It furthers
the University's objective of excellence in research, scholarship, and education by
publishing worldwide. Oxford is a registered trade mark of Oxford University Press
in the UK and in certain other countries

Published in the United States of America by Oxford University Press
198 Madison Avenue, New York, NY 10016, United States of America

© Oxford University Press 2015

All rights reserved. No part of this publication may be reproduced, stored
in a retrieval system, or transmitted, in any form or by any means, without the
prior permission in writing of Oxford University Press, or as expressly permitted
by law, by license, or under terms agreed with the appropriate reproduction rights
organization. Inquiries concerning reproduction outside the scope of the
above should be sent to the Rights Department, Oxford University Press,
at the address above.

You must not circulate this work in any other form
and you must impose this same condition on any acquirer

Cataloging-in-Publication data is on file at the Library of Congress

9780199996773

9 8 7 6 5 4 3 2 1

Printed in the United States of America on acid-free paper

To the late Larry Martin
gone too soon

# CONTENTS

*Acknowledgments*     ix
*Note on Sources*     xi

1. Can't Tell the Players without a Scorecard     1
2. Theme and Variations: Similarities and Differences among Nature's Flyers     21
3. How to Fly?     39
4. Gliding Animals: Flight without Power     60
5. Insects: First to Fly     74
6. Birds: The Feathered Flyers     103
7. Bats: Wings in the Dark     130
8. Pterosaurs: Bygone Dragons     147
9. Pedestrians Descended from Flyers: Loss of Flight     164
10. Unifying Themes?     173

*Bibliography*     183
*Index*     197

# ACKNOWLEDGMENTS

This book had an amazingly long gestation. I wrote the first draft of the first chapter in 2002. Between then and now, I talked to many people about the evolution of flight, and while I may have forgotten some of the conversations or people, they nevertheless helped shape my thinking on the topic.

Many of these conversations were with my colleagues at the University of Kansas, including Ron Barrett-Gonzalez, Dave Burnham, George Byers, Amanda Falk, Rudolf Jander, Matt Jones, and Bob Timm, as well as the many students in my Animal Flight Seminar over the years. Ed Wiley, Bruce Lieberman, and Michael Engel all helped this non-systematist understand (and keep from embarrassing myself with) phylogenetic systematics. Mark Robbins showed me museum specimens when I needed to see details of bird anatomy and feather structure and answered many ornithological questions. I learned a lot about paleontology, dinosaurs, and birds from the many hours I spent in the lab of the late Larry Martin; no one enjoyed a good argument or took it less personally than Larry. Larry's former student, Chris Bennett, patiently and thoroughly answered my many e-mails with questions about pterosaurs. I also discussed animal flight or flight evolution at various times with Roy Beckemeyer, Kristin Bishop, Sankar Chatterjee, Jeff Dawson and his grad student Ryan Chlebak, Robert Dudley, Jimmy McGuire, the late John McMasters, Jake Socha, Sharon Swartz, Jim Usherwood, and Steve Yanoviak, and there were probably others whom I am sorry to say I have forgotten. Chris Bennett, Robert Dudley, Nick Longrich, and Jake Socha each read and commented on a chapter, and comments from two anonymous reviewers who heroically read the entire manuscript significantly improved the final version. Any mistakes or inaccuracies that may remain are, of course, my responsibility. Sara Taliaferro drew most of the illustrations; hers say "(courtesy of S. T.)" in the figure legend. Also, Roy Beckemeyer and my son, Kevin Alexander, each drew a figure. (If no credit is listed with a figure, it is my own original drawing.) The staff at the University of Kansas Writing Center (especially Amanda Hemmingsen) were the first to read the drafts of my chapters, and I thank them for their useful feedback.

My biggest thanks go to my wife, Helen Alexander, for all her support and encouragement; she remains my most enthusiastic and tireless booster.

# NOTE ON SOURCES

Scientists traditionally use the "author, year" method of citing sources, which is fine for technical literature, but I have always felt that it breaks up the flow of text for non-scientists. I have chosen to use a simplified form of endnotes to cite my sources: the superscript numbers in the text refer to a set of abbreviated bibliographic entries placed at the end of each chapter under "References." The reader can then refer to the Bibliography at the end of the text for full bibliographic information for all the sources cited.

# On the Wing

CHAPTER 1

# Can't Tell the Players without a Scorecard

I was inspecting my tomato plants when I heard a familiar buzz. Looking up, I expected to see the fuzzy yellow-and-black of a bumblebee; instead, I caught a glimpse of bright metallic green. My buzzing visitor was actually a beetle, a singularly handsome scarab in a group called the bumble flower beetles. I was able to get a very nice look at my little visitor as it came to a near-standstill to inspect the foliage. The beetle hovered near a plant, probably looking for flowers for a quick bite of pollen, effortlessly matching the swaying of leaves and stems in the breeze. This beetle had a flattened, green metallic body with a coppery stripe down each side, making it *Cotinus nitida*, the green June beetle.* I could only see the beetle's wings at certain angles where the barely visible blur told me they were a good bit longer than the beetle's body. Although I know that the beetle was beating its wings at about 80 beats per second and that it uses ingenious origami-like folds to get its big wings stowed under their covers after landing, my real delight came in watching the little beast delicately inspect a couple of plants while giving me an expert demonstration of its hummingbird-like hovering abilities. After a few seconds, apparently deciding my garden offered too few rewards, the beetle rose up and shot across the lawn so fast that I lost sight of it after 3 or 4 meters (10 or 12 feet).

---

* These flashy scarabs are not all that closely related to the more common, brown June beetle. Green June beetles are bigger, more agile, and fly during the day, unlike common June beetles, which tend to be nocturnal. Scarabs make up a huge family of beetles, and biologists put green June beetles and common June beetles in entirely separate subfamilies.

What does a brief encounter with a garden insect have to do with the evolution of flight? That green June beetle represents the beetle lineage, the most successful lineage of all living animals, and its success is due in no small part to beetles' highly developed flight ability. That beetle could hover, match its movements to plants dancing in a breeze, and fly very fast. These abilities require a sophisticated and specialized flight control system—brain and nervous system, reflexes, senses—powerful muscles; exceptionally light, strong wings; and a body strong enough to withstand flight forces yet light enough to minimize the efforts of the flight muscles. Beetles arrived at their modern form through a 300-million-year period of evolutionary trial and error and continuous refinement. Beetles or robins or bats seem perfectly at home on the wing, yet their flying ability was produced over millions of generations and is still being fine-tuned today, by natural selection.

Animals have evolved flapping (powered) flight only four times in the 400-million-year history of land animals. Although flying animals cover a huge size range, wing flapping looks surprisingly similar in small insects and large birds. The basic flapping pattern does not change much because these animals all operate under the same constraints. First, they depend on the physical properties of the air; air properties do change a bit with size but are fundamentally the same for all animals. Second, all animals power their wings with muscles. Muscle properties do not change much from one animal to another, even between animals as different as dragonflies and swans, so the properties of muscles themselves determine how muscles can be employed.

## FOCUS ON POWERED FLIGHT

In the strict technical sense, flight means using a wing-like surface to produce enough lift to move through the air. A careful reader might notice that this definition includes gliding, or unpowered flight. We will certainly look at gliding because many scientists think that gliding may have been a step in the evolution of flight in one or more animal groups. A startling variety of animals can glide, even some species of frogs and snakes. Gliding is inherently limiting, however, so our main focus is on animals that use powered flight. These are easy to distinguish from gliders because they flap their wings and can stay aloft as long as their muscles permit.

The difference between gliding and powered flight is exactly analogous to coasting and pedaling on a bicycle. Coasting—like gliding—is great for going downhill but not much use for moving uphill or on long level stretches. Pedaling—like powered flight—gives true versatility. Just as a bicyclist can pedal pretty much anywhere wheels can go, a flyer can go pretty much any direction through the air by flapping. Gliding is an important part of the story, but most of my emphasis in this book is on powered, or flapping, flight.

## BENEFITS OF POWERED FLIGHT

Flying animals enjoy a number of benefits over non-flying, ground-bound (terrestrial) animals of similar size. Some are obvious. Flight gives an animal a very potent predator-escape route, one that cannot be followed by a running predator. The converse is equally true: a flying predator can attack prey from the third dimension whereas its terrestrial prey can only seek to escape in the other two dimensions. These advantages have huge potential payoffs. A flying predator might get a meal or a flying prey animal might live another day, where a non-flying animal might not.

Another clear advantage is that flying animals can reach remote or inaccessible locations that may be difficult or impossible for non-flyers to reach. The vast rookeries of seabirds on seaside cliffs of remote islands are well-known examples of this advantage. Flyers tend to reach islands or newly opened (disturbed) habitat before other animals. For example, when researchers first ventured into the devastated area around Mt. St. Helens within days after the eruption, they observed that flies were already there, long before any other new immigrants.[1]

The physics of flight requires flying animals to fly on the order of 10 times faster than a similarly sized runner can run. This speed gives flying animals yet another advantage: they can search much more effectively. Not only can a flyer cover distance 10 times faster than a runner, but the elevated vantage point of flight means the flyer can usually see much farther to each side than a searcher restricted to the ground. Thus, whether searching for food, a mate, or shelter, a flyer can search much more rapidly and cover much more ground than a runner, all else being equal.

A final advantage enjoyed by flying animals is economical travel. A flying animal uses substantially less energy (food, fuel) to travel a given distance than a runner of similar size.[2] This difference has enormous ecological consequences, for both short-term and long-term journeys. In the short term, flight makes practical daily commutes for foraging of a couple kilometers (a mile or more) for a honeybee, or many kilometers (several miles) for a great blue heron. In the long term, flight makes long-distance migration practical. Because of the way metabolic power scales with size, only large terrestrial mammals (such as caribou or African antelope) migrate significant distances, and those 300- to 500-kilometer (200- to 300-mile) migrations on foot pale next to migrations of flyers. Animals as small as butterflies and as large as whooping cranes routinely make seasonal migrations of more than 3,200 kilometers (2,000 miles). Without flight, animals could not migrate these huge distances so we would no longer see such classic seasonal indicators as wedges of geese flying south in the winter or the return of the first robins of spring.

Insects include more known species than all other animals combined. Among vertebrates, birds include almost twice as many known species as either mammals, reptiles, or amphibians. Given that most insects and birds fly, many biologists have assumed that flight played a major role in the success and diversity of these groups. Demonstrating a causal relationship between flight and evolutionary success is, however, extremely difficult. To test the question experimentally, researchers would have to go back 400 million years and re-start insect or bird evolution without flight, an obvious impossibility. Moreover, we have the example of the pterosaurs to prove that the ability to fly is not enough of an advantage to prevent extinction. Nevertheless, flight has such clear and conspicuous benefits that the ability to fly was probably an important ingredient in the success of the earliest flying species in each of the lineages that evolved powered flight.

**THE "BIG FOUR" POWERED FLYERS**

Only four animal groups have evolved the ability to fly under power, and each of these groups evolved powered flight only once. These four groups are, in the order that they took to the air, insects, pterosaurs (also called pterodactyls), birds, and bats. Although they all fly by flapping a set (or two) of wings, their basic body plan, and especially their wing structure, is quite varied. Insects are arthropods, meaning that they have a rigid exoskeleton that they must periodically shed or "molt" in order to grow. They have no backbone—nor, indeed, bone of any kind; hence, they are invertebrates. Exoskeletons are rather handy, combining the supporting framework of a skeleton with the protection of a suit of armor. They afford better protection when small, and since exoskeletons become disproportionately heavy and unwieldy if they get too large, most insects are small.

In contrast, vertebrates have endoskeletons made of bone, which are more effective for supporting a body at moderate and large sizes, so the vertebrate flyers are generally larger than insects. Insects evolved flight over 350 million years ago, and pterosaurs, the first vertebrate flyers, took to the skies over 200 million years ago, then went extinct with the dinosaurs about 65 million years ago. Birds arose around 160 million years ago, and bats were the last to evolve powered flight, more than 55 million years ago.[3] These four groups each achieved flight independently; none gave rise to any of the other groups.

## FEATS OF FLYERS, FROM HUMDRUM TO HEROIC

Flying animals are nearly everywhere, from the pigeons of many big cities to the mosquitoes of the high Arctic; from the albatrosses of the open ocean to the parrots and damselflies of the rain forest. Flying animals have discovered and made themselves at home pretty much everywhere that has air to breathe and food to eat. Mostly, they go about their business doing what all animals do: looking for food and shelter, searching for mates, and, if they are fortunate, reproducing. Flying animals have one huge advantage over those of us stuck on the ground, though. As I mentioned, flyers can fly much faster than terrestrial animals can walk or run. The very physics of flight requires flyers to move five to twenty times faster than a walker or runner of the same size can travel. For example, a small dog might be able to trot for long distances at 6 or 8 kilometers per hour (4 to 5 miles per hour), whereas a goose of the same weight can easily cruise at over 80 kilometers (50 miles) per hour. A migrating lemming would be hard-pressed to average over 2 or 3 kilometers per hour (1 to 2 miles per hour), but a pigeon typically flies at over 40 kilometers per hour (25 miles per hour). Even small flyers—a dragonfly or a sparrow, for instance—can fly much faster than I can run, and most can keep it up for hours. So a flyer can cover much more territory, and cover it faster, in the never-ending search for food or mates.

### How Fast?

Now that even the humblest flyers have left us in the dust, what about the extremes? Size plays a major role in flight speeds, as we shall see in Chapter 3, so most of the fast cruisers are big; pelicans and golden eagles clock in at around 50 kilometers (31 miles) per hour, albatrosses and swans at

60 to 70 kilometers per hour (37 to 44 miles per hour), and Canada geese at over 80 kilometers per hour (50 miles per hour).[4] The true royalty of high-speed flight, however, are the falcons. Although their cruising speeds are only 40 to 50 kilometers per hour (25 to 30 miles per hour), peregrine falcons attack other birds in flight by diving on them at high speed. These steep dives, or "stoops," give the prey less time to escape, and the falcon usually hits its hapless victim so hard that it breaks the prey's neck. Trained falcons reached speeds of 130 to 140 kilometers per hour (81 to 86 miles per hour) in dives carefully monitored by scientists,[5] and the most reliable estimates for wild falcons give diving speeds in excess of 200 kilometers per hour (125 miles per hour). At the other end of the scale, tiny fruit flies or mosquitoes can still cruise at the equivalent of a brisk walk for us, 3 or 4 kilometers per hour (2 or 3 miles per hour).[6] The ultimate low-speed extreme, of course, is hovering, something most insects and small birds can do.

### How High?

How high do animals fly? For cruising in search of food, most fly just high enough to get a good view. Worker honeybees, for instance, when searching for flowers or returning to the hive, have no need to fly higher than 3 or 4 meters above the terrain. A great blue heron commuting from its nest in a woodlot to a nearby river might fly 100 meters (330 feet) above the ground, and a nighthawk chasing insects in warm, rising air might cruise 100 to 200 meters (330 to 650 feet) above the ground. A researcher followed a flock of migrating cranes in a small airplane and saw them fly at altitudes ranging from 150 to 1,000 meters (500 to 3,300 feet).[7] Migrating birds often fly much higher, both to take advantage of favorable winds at different altitudes and to ensure that they clear any obstacles (hills, mountains) in their paths. Typical migratory altitudes for birds are anywhere from 200 to 1,000 meters (660 to 3,300 feet) above the terrain, but some rise much higher. Radar operators have tracked flocks of migrating warblers and plovers over the Atlantic at between 1,000 and 5,000 meters (3,300 and 16,000 feet), and pilots have reported eagles at 3,000 meters (9,800 feet) and sandpipers at 4,000 meters (13,000 feet). Condors soar at 6,000 meters (20,000 feet) in the Andes, and bar-headed geese and curlew sandpipers migrate over the Himalayas at 7,000 to 9,000 meters (23,000 to 30,000 feet) above sea level.[8,9] Finally, airline pilots have reported flocks of migrating swans at over 8,000 meters—over 5 miles high. Although we humans have built vehicles that fly to the outer edges of the atmosphere, consider this: for eight years after their first flight, the Wright brothers only flew as high as they felt they could safely fall, typically 10 or 12 meters (30 or 40 feet) above the ground. About eight years after that first flight, the Wrights and others finally started working to set altitude records. Although a mere eyeblink in evolutionary time, for a significant chunk of our century of human flight, people flew no higher than a three- or four-story building.

### How Far?

Pick a distance and some flyer probably forages or migrates that far. At the low end, male scale insects may just fly from the branch where they hatched to another branch on the same tree or on the nearest neighboring tree to find a mate. Certain milkweed bugs on islands in the Baltic Sea literally migrate a few hundred meters from one side of the island to the other. Most flyers, however, make more impressive trips. Foraging worker honeybees may fly 3- or 4-kilometer (2- to 2½-mile) round trips collecting pollen and nectar. That distance may not sound terribly remarkable, but scaled to body size, it would be like me making a trip (with no map, compass, or vehicle) of over 150 kilometers (90 miles), or about the distance from Chicago, Illinois, to Milwaukee, Wisconsin. And the honeybee makes several such trips each day.

On a more dramatic absolute scale, migrations include the pinnacles of animal distance records. Many birds migrate thousands of kilometers, and a substantial portion of them do so with few or no stops along the way. Songbirds and shorebirds migrate nonstop over the North Atlantic from eastern Canada and the northeastern United States to Caribbean islands and Venezuela. Their journeys range from 3,000 to 4,000 kilometers (1,800 to 2,500 miles), depending on their destination. Brant geese migrate from Alaska to Hawaii nonstop, a distance of about 4,000 kilometers (2,500 miles). One species of forest-roosting bat migrates about 1,400 kilometers (870 miles) between upstate New York and Georgia. Even insects migrate long distances. The best-known insect migrant is the Monarch butterfly, which migrates up to 3,000 kilometers (1,900 miles) from the northern United States to mountains in Mexico. Some seabirds win the long-distance prize, however. The gull-like Manx shearwater migrates 10,000 kilometers (6,000 miles) between northern European islands and Brazil. Several waders, including the curlew sandpiper, also migrate along routes of approximately 10,000 kilometers (6,000 miles) each way. The undisputed champion of long-distance travel is the diminutive Arctic tern.

This pigeon-sized bird breeds above the Arctic Circle and spends its non-breeding time on the coast of Antarctica, for a one-way distance of over 20,000 kilometers (12,500 miles). Some terns make the whole trip in a year, but others may spend two years in the Antarctic before returning north.[3]

**How Small or Large?**

Several kinds of nearly microscopic insects make up the smallest flyers. These include thrips, certain parasitic wasps, and a family of minuscule beetles. Because thrips are so tiny—from 3 or 4 millimeters (less than 2/10 inch) down to ½ millimeter (2/100 inch) in length—few people other than entomologists ever notice them. Thrips, however, are actually quite common; they can often be found in large numbers on a single dandelion flower. All the tiny insects in this size range have in common a rather strange wing structure. Their wings look more like bristly rods or loose feathers than insect wings (Fig. 1.1).[10] Because air seems more viscous to tiny creatures, these bristle-wings are just as effective at very small sizes and low speeds as more conventional wings are at larger scales.

You might wonder how big flying animals can get. This question actually has several answers, depending on exactly how you ask it. Among living animals, wandering albatrosses have the longest wingspan, almost

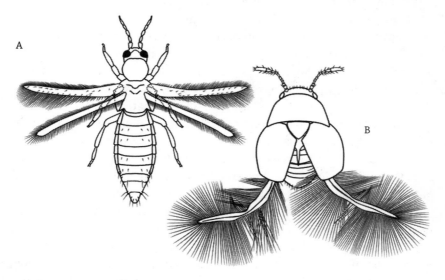

**Figure 1.1:**
Wings of very tiny insects. A. Thrips. B. Featherwing beetle. (Courtesy of S. T.)

3½ meters (12 feet). These are large birds, with a body mass of about 9 kilograms or 20 pounds (compared with 5 kilograms [11 pounds] for a bald eagle or 6 kilograms [13 pounds] for a Canada goose). Andean condors are heavier than albatrosses, at about 12 to 15 kilograms (26 to 33 pounds). Condors have slightly shorter but much broader wings. So albatrosses have the greatest wingspans, but condors are the heaviest flying birds alive today (Fig. 1.2).* If the question is restated as, "What is the largest bird that ever flew?" the prize goes to an extinct group of vulture-like birds called teratorns. Although they apparently looked and flew like vultures, teratorns were probably predators rather than scavengers. They ranged from slightly larger than condors (from fossils found in the La Brea tar pits in California) to the enormous South American species, *Argentavis magnificens*. This last species was stunningly huge, with an estimated body mass of over 75 kilograms (165 pounds; more than me!), and a wingspan of approximately 6 to 8 meters (20 to 26 feet).[11]

As big as it was, *Argentavis* was not the largest flying animal ever. The largest flying animals were pterosaurs. *Pteranodon ingens*, which lived in

**Figure 1.2:**
Very large birds. A. Condor. B. Albatross. (Courtesy of S. T.)

---

* The kory bustard, a large bird of the African plains, is often described as the largest flying bird, with typical weights of 13 kilograms (33 pounds), and a reputed maximum weight of 20 kilograms (44 pounds). The kory bustard much prefers to run and is a marginal flyer at best. It is only able to take off from level ground with difficulty and is unable to fly for more than a few seconds. It is probably less aerial than a domestic chicken, and in my opinion, not a true, fully powered flyer.

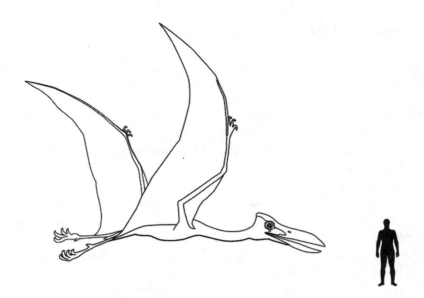

**Figure 1.3:**
*Quetzalcoatlus northropi*, the largest known flying creature. (Human figure for scale.) (Courtesy of S. T.)

what is now the central United States, was similar in size to *Argentavis*, with a wingspan of about 7 meters (23 feet). At one time *Pteranodon* was considered the largest pterosaur, but in the 1970s, it was bumped from this perch by the discovery in Texas of a partial wing skeleton of its huge relative, *Quetzalcoatlus northorpi* (see Box 1.1. Species Names). Based on size relationships of other pterosaurs, the estimated wingspan of *Quetzalcoatlus* was 11 to 12 meters (over 39 feet; see Fig. 1.3), substantially greater than a Piper Cub airplane! Weight estimates put *Quetzalcoatlus* just slightly heavier than *Argentavis*, around 80 to 90 kilograms (176 to 200 pounds).[12] So from smallest to largest, flying animals span an enormous size range, from thrips with a wingspan of less than a millimeter to pterosaurs with wingspans more than 10,000 times longer.

**Vertical Takeoff and Hovering**

In the normal course of going about their daily lives, all insects and most birds can take off vertically or near-vertically from a standing start. Moreover, as amply demonstrated by the green June beetle with whom I opened this chapter, insects and small birds can hover in one place when necessary. In contrast, relatively few of our flying machines can hover or take off

*Box 1.1:* SPECIES NAMES

In spite of a very long, complex set of rules governing the naming of animal species—the International Rules of Zoological Nomenclature—scientists occasionally manage to sneak something amusing into scientific names. For instance, *Quetzalcoatlus northropi*, the largest known pterosaur, has a bit of whimsy in its name. Before we look into that, however, we first need some background on species naming conventions.

The scientific name of a species has two parts. The first part is also the name of the genus containing that species. A "genus" is a group of closely related species. The second part is the term that identifies the particular species from among those in the genus. *Apis mellifera*, for example, is the species name of the domesticated honeybee, just one of several honeybee species in the genus *Apis*. The second term in the species name ("*mellifera*" in this case) is not capitalized and is never used alone: the species name is *Apis mellifera*, not *mellifera*. Note that genus and species names are italicized, and genus names are capitalized whether alone or part of a species name; genus and species names are traditionally based on Latin or latinized Greek words, but they can also be latinized words from other languages. Sometimes they are latinized forms of the common name—*Falco peregrinus* for the peregrine falcon—but this is rare. They can be based on any words the scientist chooses when he or she first describes the species.

The genus of *Quetzalcoatlus northropi* is named after an important deity of pre-Columbian Mexico, the feathered serpent god Quetzalcoatl (presumably because the fossils were from a flying creature found in southern Texas near Mexico). The second term, *northropi*, refers to the airplane manufacturer Northrop Company, which was well known in the mid-20th century for designing all-wing, tail-less airplanes. This pterosaur's wing was so big and its tail so small, it undoubtedly would have looked as if it were all wing when in flight.

Some species names are just fun to say, such as *Saurus soarus*, a species of gliding lizard, and *Aha ha*, a small wasp. Then there is the extinct genus of tiny, gnat-like flies, *Iyaiyai* (pronounced eye-yi-yi) and the moth species *Eubetia bigaulea* (pronounced you-betcha bygolly). *Pitohui* (pronounced pi-tooey) is the name of a genus of tropical birds with poisonous flesh and supposedly represents the spitting sound made by the scientist who discovered and tasted it. Species are often named in honor of people, such as the beetle genus *Garylarsonus*, named after *Far Side* cartoonist Gary Larson. (Larson is a favorite of biologists for his offbeat biology humor, and his name appears in several other scientific names.)

> **Box 1.1: Continued**
>
> My final example is the clever but cumbersome *Arthurdactylus conandoylensis*, a pterosaur whose fossils were found in a South American jungle. The name refers to the author Arthur Conan Doyle, writer of the Sherlock Holmes mysteries. The pterosaur name refers, not to the Holmes stories, but to a story of his called *The Lost Worlds*, in which explorers discover a living pterosaur in a jungle, capture it, and return it to London.

vertically. Those that can—helicopters, jump-jets—tend to be exceptionally complex, voraciously fuel-hungry, and much more challenging to pilot. Engineers had created useful conventional airplanes several decades before they were able to design practical helicopters, strong evidence of the great technical challenge posed by hovering and vertical takeoff (and landing).

For both vertical takeoff and hovering, a flyer needs to produce lift equal to or greater than its own weight without moving horizontally through the air. In effect, the flyer must move its wings fast enough to make up for the fact that its body is not moving horizontally. Geometry works against big flyers because as they get bigger, their weight increases with their length cubed, but their wing area and muscle force (related to muscle cross sectional area) only go up with length squared. In other words, if you double a flyer's length, its wing area and muscle force increase by fourfold but its weight increases by eightfold. Moreover, increasing weight requires wings to move faster to produce enough lift to get the flyer aloft (Chapter 3), so to hover, large flyers need to move relatively smaller wings much faster with muscles that are effectively less forceful. When animals get much bigger than small birds, this becomes a losing battle. A house fly weighs so little that it can easily move its wings fast enough to hover, a sparrow can barely move its wings fast enough to hover for short periods, and an eagle's muscles can flap fast enough for horizontal flight but nowhere near fast enough for hovering. Humans are far outside the muscle-powered hovering boundaries; we simply weigh too much relative to our muscles' abilities. A half-century of research on human-powered aircraft showed that very fit humans can barely power an airplane in level flight; hovering and vertical takeoff are effectively out of the question.

If machines that take off vertically and hover are more challenging to operate, do we see any parallels in animals? Through the action of natural selection, birds and insects have evolved an extremely sophisticated and effective flight control system. Their nervous systems and senses are finely tuned to control their flight with little or no practice or training. In fact, vertical takeoff is an important escape response for many flyers and is a hard-wired reflex in most small- to medium-sized flyers. Thus, animals don't necessarily have to learn these skills, but the specializations of their nervous systems show the level of difficulty of the task.

### Flying Both in Air and Underwater: "Amphibian" Flyers

Lots of swimming animals flap a wing-like flipper or fin to swim underwater; the basic principles of wings work just as well in the water as in air. Sea lion and turtle flippers, whale tails, and the wing-like fins of stingrays are all examples of underwater wings, or hydrofoils. Very few animals, however, can actually use the same wings to fly in air and swim under water. For example, living birds of the auk family—murres, guillemots, puffins, auklets—can "fly" in both air and water. These birds personify the "jack of all trades, master of none" approach. In order for their wings to be strong enough and small enough to work effectively under water, they are smaller and heavier than wings of other seabirds of similar size. Their wings and bodies tend to be "overbuilt" for flight so they can withstand the rigors of swimming, but compared to totally aquatic swimmers like penguins and seals, auks are light and fragile underwater.[13] Small auks tend to be better flyers and slower swimmers; in contrast, the largest auks tend to be fast, powerful swimmers, but they have difficulty getting airborne unless they can leap from a tall cliff or make a long takeoff run on calm water. (Murres, the largest living auks, sometimes migrate by swimming when the water is too rough for them to make their long takeoff run.) As we will see in Chapter 9, the recently extinct Great Auks became so large that they completely lost the power of flight in favor of strong swimming.

### From Air Combat to Aerial Sex

One of the most difficult skills for an aviator to master is air combat: intercepting and attacking other aircraft in flight. In the Second World War, the majority of fighter pilots never managed to shoot down another airplane, and roughly 10% of all fighter pilots made over half of all air-to-air kills. Even with modern tools like radar, computerized gunsights, and heat-seeking missiles, a pilot needs a great deal of skill and training to successfully intercept and engage another aircraft.[14] Contrast this with aerial predators: if a dragonfly cannot catch and eat a mosquito in the air, it starves. If a bat or a swift cannot intercept a moth on the wing, or if a falcon cannot pounce on a pigeon in flight, it goes hungry. For these aerial predators, accurately intercepting and attacking other flying animals is a matter of survival. If a fighter pilot misses too many interceptions, he or she may be assigned to some other kind of flying. If a bat misses too many interceptions, he or she will not be able to rear young and may ultimately die. The stakes are high for these aerial predators, so natural selection drove these animals to evolve the skills and instincts necessary to perform their aerial tasks.

Flying animals do an astonishing variety of things while in flight. Although I don't know of any flyers that give live birth in flight, a number of insects lay eggs in flight. For instance, some dragonflies drop them on the water surface in flight, and bee flies (whose larvae are parasites of bee larvae) eject eggs into bee burrows while hovering over them. Many beachcombers have discovered, to their chagrin, that seagulls defecate in flight. Based on radar tracking, biologists think that swifts may actually sleep in flight. I have often watched swifts and swallows drinking in flight by swooping down close to the surface of a calm pond and briefly dipping their beaks in the water to snatch a gulp, and swifts are reputed to bathe on the wing, by dragging their belly and breast feathers in the water as they skim the surface.

The most accomplished flyers even perform part or all of their mating on the wing. Many male flyers chase rivals and potential mates in the air, and they need to be able to tell which is which! Many species also have elaborate courtship rituals amounting to an aerial dance that one or both of a courting pair perform on the wing. Hawks and dragonflies are renowned for the dazzling aerobatics that males perform to attract females. Once the female shows interest, an aerial ballet involving both partners ensues. If they both manage to do the dives, climbs, turns, and twists in the proper sequence, they will copulate, and even this may occur on the wing. Damselflies, mayflies, honeybees, hover flies, and possibly some hawks all copulate in flight!

## VARIETY (AND EATING) IS THE SPICE OF LIFE

Flying animals acquire food in an unbelievable variety of ways, many of which would be difficult or impossible without flight. Osprey and fishing bats snatch fish from just below the surface while flying low over the water. Brown pelicans and terns dive into the water to catch fish, while cormorants and puffins chase fish underwater. Swifts, nighthawks, and small bats catch and eat flying insects in the air. Thrushes and starlings pluck insects from the ground and off low plants. Sandpipers and plovers use their beaks to probe into sand on beaches just above the waves for tiny invertebrates. Herons wade in shallow water and spear and eat any animal that moves. Red-tailed hawks pounce on mice, voles, and rabbits from high perches or while soaring whereas harriers do the same from a low glide over open meadows. Bluebirds, wrens, and warblers glean insects from tree branches, and woodpeckers probe under bark for insects. Tiny fig wasps actually grow up inside a developing fig, living on nourishment the fig plant provides. Adult fig wasps fly off to find another fig in which to lay eggs, and, in the process, pollinate the fig. Adult bees, wasps, and moths drink nectar from flowers (often while hovering) and bees feed their larvae pollen. Moth larvae—caterpillars—eat leaves, and in turn, many wasps feed caterpillars to their larvae. Aphids, cicadas, and some woodpeckers suck sap from plant stems and tree trunks and twigs. Termites and a variety of wood-boring beetles chew into and eat wood. Turkey vultures, burying beetles, and flesh flies eat carrion. Several kinds of beetles and flies eat dung or feed it to their larvae. Mayflies spend a year or more as aquatic, predatory larvae, and then don't eat at all as adults; they spend a day or two flying about as adults, searching for mates, mating, laying eggs, and then dying. Parasites feed on their hosts, and parasites of various stripes are not uncommon among flyers. The larvae of the human bot fly enter the skin at the site of a mosquito bite, feed on body fluids and grow just under the skin, and break through the skin to pupate as horse fly–sized adults. Frigate birds may have the most unbelievable lifestyle of all. Frigate birds are seabirds that cannot swim, so they have great difficulty fishing! A type of behavioral parasite, frigates often feed by stealing food from other birds on the wing. They attack other birds such as gannets, boobies, or gulls when the others are on their way home from a hard day's fishing. The frigate bird harasses its victim until the other bird regurgitates its fish, which the frigate catches and eats in midair. Yum!

## RECONSTRUCTING EVOLUTIONARY HISTORY: PHYLOGENETIC TREES

One of the most potent tools for studying evolution, phylogenetic systematics, was developed by biologists over the latter half of the 20th century. It has gone from being a niche technique used by a subset of taxonomists (when I was in graduate school in the early 1980s) to being a pervasive method used throughout organismal biology.[15] A modern biologist would no more try to discuss the evolutionary history of a lineage without a phylogenetic tree than she would try to measure temperature without a thermometer. So if we want to understand research on flight evolution, we'll need to know something about the products of phylogenetic systematics: phylogenies.

Phylogenies or phylogenetic trees are tree-like diagrams that scientists use to illustrate evolutionary relationships among species. In everyday terms, phylogenies represent family trees of species (or groups of species), and they illustrate ancestor-descendent relationships among species just as a family tree illustrates ancestor-descendent relationships among people in a family.

Scholars have been using tree diagrams to show various types of relationships among organisms since before Charles Darwin, and Darwin himself used what amounts to a hypothetical phylogeny to help explain natural selection in *On the Origin of Species*.[16] Evolutionary biologists and paleontologists continued to use trees to show evolutionary relationships from the late 19th century into the mid-20th century, although exactly what relationship was being illustrated and exactly how the tree was built could vary. Indeed, many such trees were based as much on the author's opinion and intuition as on hard data.

In the mid-20th century, phylogenetic systematics was introduced and it has gradually become the dominant approach to the study of evolutionary relationships. The goal of phylogenetic systematics is to study and attempt to trace patterns of evolutionary relationships, based on the fundamental assumption that all life is descended from earlier life, and modern species have thus descended from earlier species. Systematists thus seek to tease out which species are most closely related to which by comparing shared derived characteristics—features or "characters" that are shared by all descendants of a common ancestor and have been inherited from that ancestor. Traditionally, biologists have used characters from anatomy, embryology, and paleontology to assemble phylogenetic trees. Unfortunately, the vast majority of species that have ever lived are now extinct, and we only have fossils from a tiny (and biased) sample of extinct species, so paleontology is of limited help in most cases. Scientists also face the problem of determining whether characters shared by species are truly derived (inherited), versus having been evolved independently in response to similar selection pressures. For example, do swifts and swallows have long, thin, sickle-shaped wings because they inherited them from a common ancestor or because they independently evolved similar structures in response to similar selection pressures and physical constraints? The latter would be an example of convergent evolution, which happens to be more likely in the swift-swallow case. When biologists are limited to anatomical characteristics for building phylogenies, convergences and shared, derived characters can be difficult to tell apart,* which in turn can lead to ambiguities in the resulting phylogenetic trees.

Over the last decade or two, systematists have benefited from advances in both computer and molecular biological technology. The great increases in computer power coupled with huge decreases in the cost of computer memory led to quantum leaps in the number of species and the number of characters scientists could analyze on one tree. This allowed systematists to build more detailed trees and to look at ever more inclusive groups. Systematists also gained a completely new source of characters as techniques of molecular biology allowed researchers to compare similarities in DNA— the actual genetic material—among extant organisms. Such gene-based or molecular trees have gone a long way toward sorting out convergences that confounded earlier anatomically based trees. Both the traditional and the molecular approaches are in widespread use and in many cases are combined in the same study. For extinct species known only from fossils, normally the only available characters are anatomical (*Jurassic Park* not withstanding).

With four or five species, using a small handful of characters, a person may be able to assemble a phylogenetic tree in a reasonable amount of time by trial and error. This approach quickly becomes impractical as the numbers of species and characters increase. Evolutionary biologists have developed several computational (computer-based) approaches for assembling trees for large groups of species using dozens or hundreds of characters. Regardless of the complexity of the method, any systematist will always remind you that a phylogeny is a hypothesis: it is the researcher's best estimate of the patterns of descent and relationships among species. In the real world, scientists face problems that challenge even the most sophisticated tree-assembling programs. Lineages can lose characters inherited from ancestors, or characters that appear to be inherited from a common ancestor may have evolved independently in multiple lineages, or information on characters may simply be missing for various practical reasons. Moreover, when large data sets produce large, complex trees, the data often

---

* Systematists usually call characters that are similar due to convergence "homoplasies," to distinguish them from "homologies," which are characters that are similar due to inheritance from a shared ancestor that had the character.

fit equally well on more than one tree, so scientists have developed statistical methods to merge trees or choose trees most likely to be closest to the actual pattern. Lacking some way to look back in time, however, we can never know the actual patterns. Thus, even trees based on huge numbers of characters are still only estimates of the actual evolutionary pattern.

Figure 1.4 shows some simplified phylogenetic trees to illustrate some of their important features. First, the branch points represent ancestral species that diverged to form new ones; they may be known (as from fossils), but more typically they are hypothetical or assumed. Thus, the fewer branch points between two species on the tree, the more closely they share

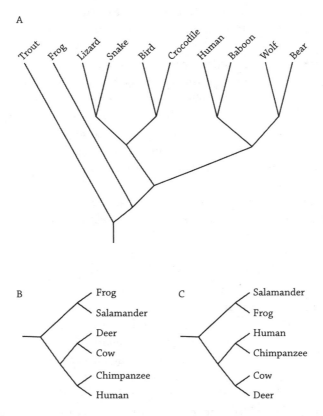

**Figure 1.4:**
Sample phylogenies or phylogenetic trees. A. Tree oriented vertically, so downward represents going back in time. Each fork or "node" represents a species (usually implied or hypothetical) ancestral to the species or nodes above. The "root" of the tree at the very base represents the common ancestor of all the species on the tree. Species (or groups of species) that are each other's closest relatives are called "sisters"; on this tree wolves and bears would be sister groups. B. Phylogenies are often rotated sideways to save space; in this case, leftward represents back in time. C. Because phylogenetic trees can rotate freely about any node, this tree is exactly equivalent to tree B. (Courtesy of S. T.)

a common ancestor. Wolves and bears share a closer common ancestor than humans and bears. Lizards share a fairly recent common ancestor with snakes, a more distant one with crocodiles, and yet a more distant one with baboons. One subtle feature is that branches are free to swivel about the branch points or nodes. The trees shown in Figures 1.4B and 1.4C both show the exact same evolutionary relationships. Humans are traditionally shown on the far right of vertical trees, subtly yet incorrectly suggesting that we represent some sort of evolutionary pinnacle; note how Figure 1.4A avoids that by rotating humans in toward the center.

Fossils play a role in phylogenies in a couple of ways. If fossils can be included on phylogenies with extant species, they can give information about when lineages arose or split in actual geological history rather than simply relative to when other lineages split. This also helps scientists determine which characters are truly ancestral. More important for our purposes, phylogenies can be constructed entirely using fossil species, which can help explain how characters evolve. For example, if we can look at a phylogeny of possible bird ancestors and see where feathers appear on the tree, we can answer questions about whether feathers evolved as a flight adaptation or for some other function in non-flying bird ancestors.

### LESSONS FROM THE BEETLE

The green June beetle from the beginning of this chapter nicely represents the end result of evolution acting on the beetle's ancestors to improve flight. The beetle's powerful flight muscles, under the control of its specialized nervous system, allow it to hover, take off and land vertically, and fly very fast. Its wings are light enough to flap without undue effort but strong enough to carry its body weight. Its exoskeleton is strong enough to anchor those powerful flight muscles as well as acting as protective armor, yet not too heavy to carry in flight. Its small size makes hovering relatively easier and fast flight relatively harder than for a larger flyer like a goose. The evolution of these specializations—flight muscles, control systems, mechanical properties, and structures—and how they are affected by overall size are some of the major threads we will follow looking at the evolution of flight in other animals.

### REFERENCES
1. J. S. Edwards (1986) *Northwest Environmental Journal.*
2. K. Schmidt-Nielsen (1972) *Science.*

3. D. E. Alexander (2002) *Nature's Flyers: Birds, Insects, and the Biomechanics of Flight.*
4. H. Tennekes (1996) *The Simple Science of Flight: From Insects to Jumbo Jets.*
5. V. A. Tucker, T. J. Cade, and A. E. Tucker (1998) *Journal of Experimental Biology.*
6. J. H. Marden, M. R. Wolf, and K. E. Weber (1997) *Journal of Experimental Biology.*
7. C. J. Pennycuick, T. Alerstam, and B. Larsson (1979) *Ornis Scandinavica.*
8. L. W. Swan (1970) *Natural History.*
9. I. Newton (2008) *The Migration Ecology of Birds.*
10. R. F. Chapman (1982) *The Insects: Structure and Function.*
11. K. E. Campbell and E. P. Tonni (1983) *Auk.*
12. S. Chatterjee, and R. J. Templin (2004) *Geological Society of America Special Papers.*
13. D. Boag and M. Alexander (1986) *The Atlantic Puffin.*
14. J. F. Dunnigan (2003) *How to Make War: A Comprehensive Guide to Modern Warfare in the Twenty-first Century.*
15. E. O. Wiley and B. S. Lieberman (2011) *Phylogenetics: The Theory of Phylogenetic Systematics.*
16. C. Darwin (1860) *On the Origin of Species.*

CHAPTER 2

# Theme and Variations

*Similarities and Differences among Nature's Flyers*

Pigeons and dragonflies are both skilled flyers, and they both fly by flapping their wings. The flapping process is essentially the same in these two flyers (Chapter 3), but you need only a quick look to see that the structure of their wings is completely different. The pigeon's wing surface is made mostly of an array of dozens of partly overlapping feathers, supported by a highly modified front leg; the dragonfly wing uses a complex arrangement of rigid struts—called veins—supporting a thin, flexible membrane. Birds and insects followed very different evolutionary routes to achieve flight. Although their wings do the same thing, natural selection worked with different raw material in evolving birds' and insects' wings. What structures evolved into wings, and what did the animal use those structures for before evolving wings? These are among the most important questions in the evolution of animal flight. They have inspired many vociferous arguments, and even today we have only partial answers to them.

Natural selection, the process that leads to most evolutionary change, can work only with the materials and structures that an organism already possesses. As a potential flying animal, I might have more effective wings if I could make them out of fiberglass or aluminum, but if I am a vertebrate animal, I am stuck with using (and modifying) things like skin, bone, muscle, and tendons. Likewise, if I were an insect, I might be able to make stronger wing veins if I could use a graphite-epoxy composite, but I am limited to using the material that makes up my exoskeleton. Natural selection works by gradually modifying what an animal already has, not by providing totally novel materials or structures. Biologists call these limitations "historical constraints." If I am an insect, I am unlikely to evolve feathers because my outer body covering, an exoskeleton, is non-living and does not grow. Moreover, my exoskeleton is mostly made up of chitin, a very different substance from keratin, the material that makes up a bird's feathers (keratin is a major component of vertebrate skin but not found in insects). These are historical constraints; my ancestry limits how much natural selection can change my lineage.

### THE KEY: FUNCTION OF PROTOWINGS

In order to evolve a wing in the face of historical constraints, natural selection must have acted on some structure that had at least slightly wing-like properties, perhaps a large surface area or something extending out on each side of the body. Scientists sometimes call such structures "protowings," but this name can be misleading because the original structure most likely had no aerodynamic function at all. Bat wings, for example, are modified front legs, and the ancestors of bats undoubtedly used them for climbing trees and walking. If natural selection, in improving the original function of the protowing, also *by coincidence* gave it a bit of aerodynamic ability, the protowing would have become a structure with two functions. Perhaps long arms and big hands for climbing also helped stabilize the creature's leaps between branches; if this were a big enough benefit, further increase in surface area for even better stabilizing might occur, allowing natural selection to act on the aerodynamic function and beginning the process of evolving true wings. The protowing would go through a transition period, when it would retain the original function while improving its flight functions, such as a front leg being used for both climbing and gliding. If gliding proved to be sufficiently beneficial, natural selection would favor improving the aerodynamics of the protowing to the point that it becomes a fully functional wing.[1] This process of evolving a new use for an existing structure and eventually losing the original function is called "exaptation." Even quite complex structures can evolve this way. (See Box 2.1. Evolution of Complex Structures.)

### MUSCLE-POWERED FLIGHT

All flying animals use the same "engine" to power their flight: muscles and muscle tissue. Muscle tissue works the same way in all animals, and

> *Box 2.1:* EVOLUTION OF COMPLEX STRUCTURES
>
> Can natural selection produce a structure as finely tuned and specialized as a flapping wing? Since the time of Charles Darwin, skeptics of evolution have raised the same objection: since a complex organ like a vertebrate eye or a mammalian kidney works only if it has all its parts, how can it have evolved gradually? If my eye is missing any of its major parts—lens, iris, retina, cornea—it doesn't work. Darwin himself responded to this objection and explained how such complex organs could have evolved gradually, and his explanation still suffices today. His key point is that the function must change as the structure becomes more complex. For instance, an eye does not have to form a sharp image to be useful (even though I would be considered "blind" if my eye does not form an image): animals as simple as jellyfish and flatworms have "eyes" that really just detect the presence or absence of light. Imagine that the skin over such an eye becomes thickened to form a lens. Now this eye is directional, so the animal can tell more precisely where the light is coming from. Such an eye might be very useful for avoiding predators, even though it does not form an image. In addition to being directional, however, lenses focus an image whether or not the light sensors are capable of distinguishing that image. Initially, light sensors would not have been able to resolve an image, but any increase in the size or resolution of the light sensors would form a retina and allow the eye to start detecting a crude image. Further elaboration of the retina to sharpen and refine the image would lead to a fully functional eye.
>
> So, yes, natural selection is up to the task of forming complex structures such as eyes, kidneys, and wings. Both the function and the structure start out simple—and possibly unrelated to later functions—and as they both are refined by natural selection, the function becomes more sophisticated right along with the structure.

muscle tissue from an eagle and from a mosquito are quite similar, down to the arrangement of protein filaments inside the cells that do the work. The key operating principle of muscle is that it works by *shortening* or *pulling*.[2] Muscles can't push or lengthen forcefully, and they can't rotate. An engineer would say that they are linear motors that work in only one direction. Muscles are usually arranged in antagonistic pairs, where each member of the pair reverses the action of the other member—in humans, the biceps muscle bends the elbow, and the triceps straightens the elbow, making them an antagonistic pair. So all flying animals have one (or more) downstroke muscle and one (or more) antagonistic upstroke muscle.

Knowing how muscles work, we can see that flapping to power flight is a natural arrangement. Antagonistic muscles pretty much require some sort of to-and-fro motion rather than a continuous rotation. All four animal groups that use powered flight independently evolved the up-and-down flapping movements because they all use muscles with the same operating properties.

Muscles provide both power and force for flapping flight. People use "power" and "force" interchangeably in everyday speaking, but in science they are quite distinct. Force is a push or pull, something that can accelerate or decelerate an object. Power is the rate of doing work, that is, work performed divided by the time to perform it. Force and power are connected by work, because work is the force applied to an object times the distance moved. (If the object does not move, then no work is done and no energy is consumed: as my former graduate advisor liked to point out, the chain holding a chandelier provides the force for free, no fuel required.) If I use a rope to lift a 1-kilogram (2.2-pound) brick 10 meters (33 feet), I must pull on the rope with a force of 9.8 newtons (2.2 pounds-force), regardless of how fast I raise the brick. If I raise it quickly, however, I use more power than if I raise it slowly. I use 10 times as much power raising it at 10 meters per second as raising it at 1 meter per second, even though I pull it with approximately the same force. In both cases, I have done the same amount of work—9.8 newtons times 10 meters, or 98 joules of work—but in the first case I have done the work 10 times as fast, so I have used 10 times as much power.

Wing physics dictates that animals must move their wings very fast for them to be effective, so power is important. As a general rule, flight requires more power than walking because the flapping movements must be much faster.[3] Curiously, the amount of force the muscles must provide need not be much different from walking or running, so the muscles don't need to be unusually strong, but they do need to be fast. So, to supply the needed force while moving very rapidly, wing muscles must be powerful due to speed rather than force.

The high power requirements of flight run up against another feature of the way muscles operate. Muscle tissue is specialized for either long periods of continuous use or brief bursts of intense activity. The "continuous use" type is called "aerobic" because it requires a continuous supply of oxygen while it is active. The "burst" type is "anaerobic," meaning that it does not require oxygen during its bursts of activity.[4] Why the difference? Anaerobic muscle tissue can react faster and it is usually stronger—it can lift a heavier load than the same amount of aerobic muscle—but it produces lactic acid as a byproduct during activity.[5]

Anaerobic muscles can react faster partly because they run on their own stored fuel (carbohydrates), but once this fuel is used up, they must rest while the body replenishes their fuel. Moreover, when lactic acid builds up to high levels, the muscles cannot contract any more until the lactic acid is recycled or removed. Aerobic muscles, in contrast, are not as strong, but they can continue to contract as long as the body supplies them with oxygen and fuel (usually fats). In vertebrates, the difference in muscle tissue is easily visible: aerobic muscles contain an oxygen storage pigment, myoglobin, that gives them a deep red color. Anaerobic muscles lack myoglobin and tend to be pale pink or white. This difference is the basis for "dark" and "white" meat in chicken: dark meat is aerobic muscle and white meat is anaerobic. Chickens spend almost all their time on the ground and fly only in emergencies, so their leg muscles are aerobic and their flight (breast) muscles are anaerobic. In contrast, most ducks, which are powerful long-distance flyers, have dark breast meat—their flight muscles are aerobic.* To see the tradeoff and understand the consequences for an animal's body, think of aerobic muscles working on a "pay as you go" basis. They must receive a continuous supply of oxygen and fuel, but they produce little or no lactic acid. Anaerobic muscles, however, work on credit. They can work quickly and forcefully, but they produce a toxic byproduct, lactic acid, that the body recycles via oxygen-consuming reactions; biologists even call the lactic acid buildup the "oxygen debt." Clearly, flying any great distance or for any substantial duration will depend on aerobic flight muscles. Some of the controversy over birds' ancestors (Chapter 6) revolves around whether the ancestors of birds (or the earliest flying birds themselves) had aerobic flight muscles.

## EVOLVING EXPERTISE
### Brains and Sense Organs

The flapping movements of powered flight are much more than a simple up-and-down motion (as we will see in Chapter 3). To fly effectively, animals had to evolve nervous systems that could produce the appropriate movement patterns and reflexes, not only to fly straight and level but to

---

* Our muscles, like those of other mammals, usually contain both types of cells, although most muscles tend to have substantially more of one type or the other. Just to add confusion, we mammals even have a couple of intermediate cell types that have properties in between aerobic and anaerobic muscle cells.

maneuver as well. Flying crows or bumblebees don't use their tails like boat rudders to turn; they change direction by modifying their wingbeat pattern. These wing movements would have been quite different from walking or climbing, and they were not automatic in the beginning. Natural selection would have modified the nervous system's "walking" control patterns to allow them at first to control both walking and flying and later to primarily control flight.

Sense organs would have changed as well. Flyers tend to fly as much as 10 or 15 times faster than a terrestrial animal of similar size can run. So flyers need faster and more precise sensing to avoid obstacles than their ground-bound relatives. Flyers tend to have better vision than similar non-flying animals, for example. Moreover, flyers need to sense new things: is the air flowing smoothly over my wings? Am I rising or sinking? Clearly, to get the most out of flight, flying animals need to evolve different sensing abilities. Early on, such sensing (and the associated control systems) would have been crude and rudimentary, but selection for more effective flight mechanisms produced the highly specialized flyers present today.

### Stability and Maneuverability

Modern airplanes tend to be either "stability configured" or "control configured."[6] Stability-configured airplanes tend to fly straight and level by themselves; if knocked off kilter by a gust or turbulence, they tend to right themselves automatically, just due to their physical arrangement. Control-configured airplanes don't have built-in stability; they tend to change direction or orientation with the slightest gust. Strongly control-configured airplanes can be so unstable that they require some sort of automatic-stabilizing control system to actively tame their lack of stability and make them flyable.

Why have one versus the other? Stability-configured airplanes are easier and safer to fly, are simpler to build, and may be more economical to operate (because they tend to fly straighter, a straight line being the shortest distance between two points). Training and most transport and passenger airplanes are normally stability configured. Because they are so stable, these craft tend to resist steering efforts, so they are not very agile. Conversely, the main advantage of control-configured airplanes is that they are extremely agile and maneuverable. Military combat aircraft, especially fighters, are often control configured. In combat, maneuverability outweighs economical operation.

Modern flying animals are, by and large, control configured. They must continuously monitor and adjust their flight path, because they have limited built-in stability.[7] They have what amounts to an automatic stabilizing system of sensors and reflexes built into their nervous systems. This arrangement gives flying animals unparalleled maneuverability. Humans have yet to build a flying machine with half the aerial agility of a robin or house fly.

Over half a century ago, aeronautical-engineer-turned-evolutionary-biologist John Maynard Smith noticed the extreme maneuverability of modern flying animals and realized that it represents a potent specialization. He pointed out that the earliest member of any lineage to fly could not have had such specialized abilities; it would have needed a physical design with more built-in stability.[8] He went further and pointed out that we should expect primitive flyers to be inherently stable; conversely, highly maneuverable, inherently unstable flyers with active stabilizing mechanisms must therefore be more derived (less primitive). In a sense, Smith's theory is hard to demonstrate, because all modern flying animals are descendants of lineages that have been flying for tens of millions of years. His concept turns out to be helpful, however, in explaining potentially stabilizing features of very primitive flyers like *Archaeopteryx* and early pterosaurs (as we will see in Chapter 6).

## OTHER DESIGN TRADE-OFFS

Modern flying animals have largely evolved into control-configured flyers with active stabilization. Other design trade-offs have not been so clear-cut. (See Box 2.2. Concept of "Design" in Evolution.) The trade-off between structural strength and lightness, for example, depends on the animal's size, among many other aspects of its biology. Sometimes anatomical structures don't scale up or down very well. Moths and mosquitoes, like all insects, have exquisitely light, somewhat umbrella-like wings that are surprisingly strong for their weight. Scaled up to the size of a crow's wing, however, they would be heavy and too weak to carry the crow's weight (Chapter 3). The crow needs a lighter, stronger wing, and, like all birds, its wings are of a completely different design from an insect's, based on a modified front leg and feathers. Even among birds, some have wings that are more lightly built while others have more robust wings. Gulls have long, lightly built wings suitable for flying and soaring in wide-open spaces, whereas puffins have short, strong, heavy wings because they use their wings to swim underwater as well as to fly.[3,9]

> **Box 2.2: CONCEPT OF "DESIGN" IN EVOLUTION**
>
> Biologists, as a rule, tend to avoid using the term "design" when applied to living organisms. This reluctance is not because the concept of an "animal's design" is somehow invalid but because they fear that some people may take it to imply the work of a conscious designer. Such self-censorship is unfortunate because when we compare different structural features of, say, a bird wing and a bat wing, which have important and obvious functional consequences, we are really comparing two different wing designs. The overwhelming majority of biologists take it as given that these two different wing forms resulted from evolution operating through natural selection. When I, as a biologist, mention the "design" of some feature of an animal, I am simply referring to the overall structural arrangement, usually implying various functional advantages and disadvantages. I am not implying that a deity or supernatural designer is necessary, just that the animal has a particular structural arrangement different from other animals. The "designer" is natural selection.

As specialized as flying animals are, sometimes processes other than flight affect their weight. Fat storage for migration or eggs about to be laid can greatly increase a flyer's weight. This weight increase requires stronger (and heavier) wings in order to carry flight loads. If animals evolve wings strong enough for periods when their bodies are unusually heavy, their wings might appear to be "overbuilt" at other times.

**Speed versus Distance**

Fast flight speeds trade off against flight distances (or duration, which amounts to nearly the same thing). High flight speeds are clearly useful in some situations, such as catching prey or avoiding predators, for example, or covering lots of territory quickly while searching for food or mates. High flight speeds, however, require disproportionately more power, meaning that a fast flyer runs out of fuel sooner and covers less total distance. Lower speeds are more economical, so animals flying long distances tend to fly a good bit slower than their maximum speed. Moreover, flight forces are greater when an animal is flying fast, so it needs a stronger body structure to fly very fast. A more leisurely flyer can get by with a lighter structure. The choice depends on the animal's lifestyle. For their size, ducks and falcons and hover flies fly fast, butterflies and gulls and vultures fly slowly.[10]

**Drag Reduction**

Any object moving through the air experiences drag, a slowing force or a force that tends to decelerate the object. Most flyers would benefit aerodynamically from reducing drag as much as possible. For example, fast flyers could fly faster on the same power, or long-distance flyers could fly farther on the same fuel load, if their drag was lower. What can natural selection do to reduce flyers' drag?

For larger animals—most birds and bats, and the largest insects—streamlining and smoothing the body surface reduces drag. For small- and medium-sized insects (mosquitoes or house flies, for example) streamlining is less effective, and being more spherical to reduce surface area is more effective (Chapter 3).

For some animals, drag reduction is of paramount importance. Fast-flying falcons are highly streamlined; so are gulls and terns, which often fly very long distances. Turkeys, on the other hand, are not nearly so streamlined. Turkeys walk a lot more than they fly, so drag reduction is of negligible benefit to them. Or consider mosquitoes: neither streamlined nor spherical, they have big feathery antennae, long legs, and roughly cylindrical abdomens. The ability to locate and walk on hairy prey—big antennae and long legs—and swell up with a big blood meal—cylindrical abdomens—clearly take precedence over drag reduction in these creatures. In a similar way, the male pheasant's long tail or the shape modifications that allow a leafhopper bug to look like a thorn or a katydid to look like a leaf surely incur drag penalties. The benefits, however—the pheasant attracting mates, the insects hiding from predators—in these cases outweigh the drag costs.

## HOW TO BUILD A WING: FOUR VARIATIONS ON A THEME

In the next chapter, we will look in detail at some of the aerodynamic characteristics of effective wings. For now, let us assume that a wing needs to be a large, light, more-or-less flat surface with a good bit more surface area than a typical leg. Its performance will improve some if it is streamlined in cross section, that is, a bit thicker in front than at the back. It will also perform better if it is a bit "cambered," meaning the top is convex or slightly bowed upward. The wing needs to be able to flap, which involves rather complicated up/down, fore/aft, and twisting movements. So it needs a joint or articulation with the body that allows it to make these complex movements while still supporting the body's weight in flight. Some ability to change the wing's shape or area is also handy, because different circumstances may require adjusting the wing. For instance, birds greatly reduce their wing area during steep dives compared with level flight. An animal's wing has other, non-aerodynamic requirements as well. It usually needs to fold up in some compact way so that the animal can get around when not flying, for example. These are the primary requirements of a flapping wing, although particular animals may have other, more specialized requirements.

**Bird Wings**

Of all the flying animals, the ones people are probably most familiar with are birds. Bird wings are highly modified front legs, but surprisingly little of the wing is skin, flesh, or bone. In a typical bird wing, at least 75% of the wing area is just feathers. The part of a bird's wing corresponding to our upper arm, forearm, hand, and fingers is much shortened and extends little more than half the length of the wing. The upper arm bone—humerus—and forearm bones—radius and ulna—though small, don't look all that different from ours (Fig. 2.1). The wrist and hand, however, are massively reduced, from more than a dozen bones in mine (not counting finger bones) to a mere three in a pigeon. As for fingers, birds only have three: a semi-mobile thumb with one or two bones, and a large (two bones) and a small (one bone) finger functionally fused into a single unit. The thumb carries a small group of feathers, the alula or bastard-wing, which the bird probably uses to adjust the airflow over the main wing. Although they are relatively shorter and fewer in number, the hand and finger bones of birds look quite stout compared to ours.[11]

The main flight feathers (primary and secondary feathers) are attached by ligaments directly to bones. The large primaries that make up the wingtip attach to the hand and finger bones. The secondaries, which make up the back half of most of the wing, are attached to one of the forearm bones (Fig. 2.1).

We can move the joints in our arm and hand independently, but a bird's joints, muscles, and ligaments all work together. When a bird straightens

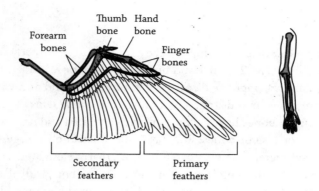

**Figure 2.1:**
Bird wing skeleton and feathers, with bones shaded. (Human arm shown for comparison, not to scale.)

(extends) its elbow, its wrist and fingers also extend; when it bends (flexes) its elbow, the other joints flex automatically as well.[11,12] Moreover, the primary and secondary flight feathers are attached to the skeleton and to each other at their bases so that when the arm and hand extend, the feathers fan out like a hand of playing cards. When the arm and hand flex, the feathers slide together in a neat stack automatically.[13] This movement illustrates the elegance of the bird wing's design: the bird can fully extend its wing or it can partly flex it to reduce its area, and the wing always stays smooth and stiff because the feathers automatically overlap each other as the wing flexes. So birds can change the size of their wings in flight much more than other flying animals. Also, losing a few flight feathers doesn't keep a bird from flying, the way a damaged wing might ground other flyers. In fact, birds replace their flight feathers annually or "molt," and most birds molt their flight feathers a few at a time and never lose the ability to fly.

Feathers are remarkable structures. They are very light for their surface area, yet stiff enough to carry flight loads. Contour feathers—the flat, stiff ones, not the fluffy down feathers—can have their side branches or "barbs" thoroughly mussed and be preened back into a smooth, stiff surface. The Polynesians' armor made from bird feathers was not just for ornamentation. Several layers of contour feathers can provide a surprising amount of protection from impacts and penetration (as birds, no doubt, discovered quite a long while before the Polynesians). Each primary feather has an airfoil-shaped cross section (Chapter 3) and in many birds, these primaries can each function as a tiny, individual wing.

### Bat Wings

The flight surface of a bat's wing, unlike a bird's wing, is directly supported by the skeleton (Fig. 2.2). Bat and bird wing skeletons, in fact, have specialized in almost opposite directions. Where bird hand and wing bones have become shorter and stouter, with lots of fusion and loss of bones, bat hand and finger bones have become greatly elongated and narrowed, with little fusion or loss of bones. The bat wing consists of a large, stretchy, multi-layered membrane supported by the greatly elongated arm, hand, and finger bones. The membrane is anchored to the body wall all along the bat's trunk and extends back along the legs to the ankles, as well as from the legs to the tail.[14]

Biologists divide the wing membrane into four regions, two small and two large ones. The propatagium or prowing is a small triangle in front of the elbow, from the shoulder to the wrist. The handwing is the membrane between the bones of the hand and fingers, which makes up approximately the outer half of the wing. The inner half is the armwing, which runs from the handwing (last finger) to the body wall and leg and makes up the inner half of the wing. Finally, the uropatagium or tailwing runs between the two hind legs and usually incorporates the tail.[15] The membrane is much more than just a top and bottom layer of skin. It contains an elaborate network of fibers, some tough and others stretchy, plus several of its own highly specialized muscles. Bats use these fibers and muscles to help control the shape of the wing in flight.[16]

Imagine taking your arm skeleton, lengthening and thinning the upper arm and forearm bones, and greatly lengthening your hand and most finger bones, while leaving your thumb and wrist bones about the same size. That is pretty much the arrangement of the bat wing skeleton. The

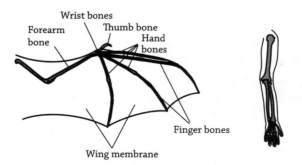

**Figure 2.2:**
Bat wing structure, showing bones and wing membrane or patagium (shading and human arm as in Figure 2.1).

clawed thumb is free of the wing membrane and relatively mobile. Bats use their thumbs for grasping and climbing. The second digit, the bat's index finger, is less than half as long as the other wing fingers and acts mainly as a stiffener of the wing's leading edge, reinforcing the third digit. The third, fourth, and fifth digits, corresponding to our middle, ring, and pinkie fingers, support most of the handwing (Fig. 2.2); their joints give the bat great ability to modify the wing's shape, particularly its curvature or camber (Chapter 3).

Bats have evolved a striking functional similarity to bird wings: bat wings also have an automatic flexing and extending mechanism. As with birds, when the bat extends its elbow, that movement also causes the wrist and hand to extend.[17] The structure is different in bats, however. In birds, the shape of the elbow joint and the parallelogram-like arrangement of the forearm bones are as important as the muscles and other soft tissue, but in bats the mechanism relies only on muscles.[15]

Although a number of books and reviews state that the armwing supports the body's weight and the handwing produces thrust, this description is a great oversimplification.[15,17] In fact, the whole wing produces upward and forward forces. The proportion of upward force is high at the base and decreases toward the tip, while the proportion of thrust is high at the tip and decreases toward the base. The change in upward and forward force along the wing is gradual, and there is no sharp division between lift- and thrust-producing regions.[3] This distribution of upward and thrust forces applies to any flapping wing, not just those of bats.

**Pterosaur Wings**

Pterosaur wings look superficially more like bat wings than bird wings. The basic pterosaur wing arrangement had a single, long, supporting skeletal arc at the front bearing a membrane stretching from the wingtip to the body wall and back to the legs (Fig. 2.3). The skeleton that formed the front of the wing had unremarkable arm bones, hand bones that were short in some species and elongated in others, and one tremendously elongated finger.* The wing finger typically was longer than the rest of the wing skeleton combined. Pterosaurs also had three normal-sized, clawed fingers separate from the wing membrane, making the elongated wing

---

* "Pterodactyl," the alternate common name for the group, literally means "wing finger." Scientists tend to prefer "pterosaur" ("winged lizard") as the common name because "pterodactyl" technically refers to one subgroup of pterosaurs (see Box 8.1).

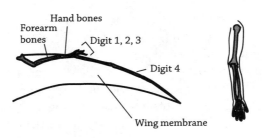

**Figure 2.3:**
Pterosaur wing structure (details as in Figure 2.2).

finger the fourth digit.[18] From fossils preserving impressions of the wing membrane, the membrane was apparently flexible enough to fold up when the animal was not flying.[19]

Scientists long thought that pterosaur wings were sort of simplified bat wings; they now think pterosaur wings were actually quite different from bat wings. For over a century, scientists envisioned the pterosaur wing membrane as a flexible sheet that would have billowed freely, with no means of controlling its shape other than the tension of the fingertip stretching the membrane away from the body. A couple of decades ago, when paleontologists first looked seriously at the mechanics of pterosaur wings, they realized that a floppy membrane would not allow the pterosaur to flex its wings to reduce area (it would go slack if flexed), and keeping the membrane stretched taut would place improbably high loads on the wing finger's slender skeleton. Back in the late 1800s, paleontologists noticed what appeared to be some sort of fiber in impressions of the wing membrane. When modern scientists looked more closely at those fibers, they concluded that the fibers were probably membrane stiffeners.[20] The fibers run in the same direction as the shafts of primary and secondary feathers of bird wings,[21] and if they were stiff like cartilage or fingernails, they would have been ideally suited to keep the membrane spread out: they would have kept the membrane from getting narrower as the pterosaur stretched out its wings. Stiff fibers would take some of the load off the finger bones and also would have allowed the pterosaur to flex (shorten) its wing in flight, to reduce the wing's surface area without allowing it to billow or go slack.[21,22]

So far, we cannot tell from the fossils if pterosaurs had an automatic wing extension-flexion mechanism like birds and bats—and because we only know them from fossils, we may never know. A close study of the joints may give some hints, but without the muscles, we won't be able to say for sure. Since the muscles themselves don't normally fossilize, scientists must reconstruct muscles based entirely on markings on the bones, and this technique is not likely to be accurate enough to answer this question.

**Insect Wings**

Unlike the wings of all other flying animals, insect wings are not modified legs; indeed, insects are the only flapping flyers that did not give up any legs in the process of evolving wings. (Just what was the precursor for insect wings is the subject of a spirited debate; see Chapter 5.) Insect wings are attached to the middle body region, or thorax, on the sides near the top; the legs are attached to the bottom of the thorax, and adult insects normally have six legs, regardless of whether they have wings.

The insect wing structure is completely different from the others we have seen. Insects are arthropods, meaning they have a rigid exoskeleton outside their body rather than an internal skeleton like us. The exoskeleton is a secreted, non-living material, in somewhat the same way that our fingernails or hair are non-living. Unlike nails or hair, however, insect exoskeletons don't grow once they have formed, and they can't be repaired or modified. In fact, insects must periodically shed their exoskeletons and secrete and harden a new one in order to grow. Insect wings form from modified regions of exoskeleton. The material of exoskeleton, called cuticle, can be hard, where it forms rigid plates or tubes, or soft, where it forms hinges or joints.[23]

The insect wing is essentially a sandwich of two layers of cuticle. Over most of the wing, these two layers are extremely thin and joined tightly together to form the wing membrane. The membrane is supported by rod-like "veins," which are also formed from the same two layers of cuticle; the veins, however, are hollow, with much thicker walls and a blood-filled core. The membrane is delicate and flexible, so the veins are the main load-bearing structures (Fig. 2.4). The veins radiate out from the wing base to the tip like spokes, and the main lengthwise veins are often connected by cross-veins.[24] The exact arrangement of the veins varies tremendously from group to group and is so specific that biologists often use the vein pattern to identify insect families and sometimes even species.

Most insect wings are not simply flat arrays of veins and membranes. The wings tend to be slightly pleated, so that some veins are on top of low ridges and others are at the bottom of shallow valleys (Fig. 2.4). These pleats are usually well developed at the front of the wing and dwindle away toward the rear. This combination of veins and pleating adds a lot of stiffness and resistance to lengthwise bending at almost no weight penalty.[25]

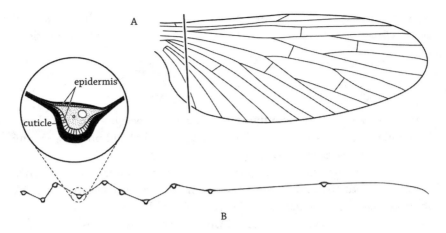

**Figure 2.4:**
Insect wing structure, showing the strengthening struts or "veins" supporting the delicate membrane. A. Top view. B. Cross section. (Courtesy of S. T.)

Moreover, scientists were surprised to discover that the pleats had little or no adverse aerodynamic effects.[26] Because of the small size of insect wings, air tends to flow smoothly over the pleats as if they formed a smooth surface from peak to peak.[26,27]

Another huge difference between insect wings and those of vertebrate flyers is that insect wings contain no muscle or actively powered joints—nothing equivalent to elbows or wrists. Insects move their wings entirely by muscles acting on the wing base at its joint or articulation with the thorax.[24] Without any muscles or joints out in the wings, insects have no way to flex the wing to reduce its area while flapping. Most insects can fold their wings over their bodies to keep them out of the way when not in use, and many insects can fold up their wings like fans or accordions to stow them between flights, but they generally cannot use these folding mechanisms to reduce wing area in flight. Moreover, because insect wings contain so little living tissue, insects cannot normally shed them when molting, which means only fully grown adults have functional wings.

Finally, where birds, bats, and pterosaurs have a single pair of wings, the basic insect pattern is to have two pairs of wings. Some insects, such as dragonflies and katydids, have more-or-less equal, independent pairs of front and hindwings. Some insects have reduced hindwings or they have mechanisms to couple the hindwings to the forewings so they function as a single pair; many, such as butterflies, bees, wasps, cicadas, and aphids, have both. Flies have modified the hindwings into sense organs, and beetles have turned the front wings into protective covers, so they each have a single functional pair of wings. Insects thus have more flexibility in number of wings than other flying animals.

## THE BIG FOUR

Over the course of animal evolution, dozens of species have evolved unpowered flight or gliding (as we will see in Chapter 4). In all that time, only four lineages have evolved powered or flapping flight: insects, pterosaurs, birds, and bats. Flapping is a much more sophisticated process than gliding, so evolving flapping seems to have required overcoming a much higher evolutionary hurdle than gliding. In later chapters, we will cover each of the Big Four groups in turn. I will describe the circumstances that scientists think may have favored flight and the features of the animals involved that may have predisposed them to evolve wings. Some of these topics remain open questions, but we will see that researchers have used indirect evidence and clever experiments to narrow down the range of possibilities.

## REFERENCES

1. S. J. Gould (1985) *Natural History*.
2. T. A. McMahon (1984) *Muscles, Reflexes, and Locomotion*.
3. D. E. Alexander (2002) *Nature's Flyers: Birds, Insects, and the Biomechanics of Flight*.
4. S. Vogel (2001) *Prime Mover: A Natural History of Muscle*.
5. K. Schmidt-Nielsen (1990) *Animal Physiology: Adaptation and Environment*.
6. M. J. Abzug and E. E. Larrabee (1997) *Airplane Stability and Control: A History of the Technologies That Made Aviation Possible*.
7. D. R. Warrick and K. P. Dial (1998) *Journal of Experimental Biology*.
8. J. Maynard Smith (1952) *Evolution*.
9. H. Tennekes (1996) *The Simple Science of Flight: From Insects to Jumbo Jets*.
10. J. M. V. Rayner (1988) *Current Ornithology*.
11. J. J. Videler (2005) *Avian Flight*.
12. H. I. Fisher (1957) *Science*.
13. A. S. King and D. Z. King (1979) in *Form and Function in Birds*.
14. C. J. Pennycuick (1972) *Animal Flight*.
15. J. D. Altringham (1996) *Bats: Biology and Behaviour*.
16. S. M. Swartz, M. S. Groves, H. D. Kim, et al. (1996) *Journal of Zoology*.
17. G. Neuweiler (2000) *The Biology of Bats*.
18. P. J. Currie (1991) *The Flying Dinosaurs*.
19. D. M. Unwin and N. N. Bakhurina (1994) *Nature*.
20. K. Padian (1991) in *Biomechanics in Evolution*.

21. K. Padian and J. M. V. Rayner (1993) *American Journal of Science*.
22. S. C. Bennett (2000) *Historical Biology*.
23. P. J. Gullan and P. S. Cranston (2010) *The Insects: An Outline of Entomology*.
24. R. F. Chapman (1982) *The Insects: Structure and Function*.
25. R. J. Wootton (1986) *Journal of Experimental Biology*.
26. C. J. C. Rees (1975) *Nature*.
27. B. G. Newman, S. B. Savage, and D. Schoulla (1977) in *Scale Effects in Animal Locomotion*.

# CHAPTER 3

# How to Fly?

Eagles fly, ducks fly, bumblebees fly, house flies fly. They all fly with wings, so before we can look at the evolution of flight we first need to know what wings are and at least a little about how they work. In the previous chapter, we looked at the wing anatomy of the flying lineages. Now we will turn to the physical properties of wings in general and see what properties make wings more or less effective. All wings, from those of house flies to those of airliners, are subject to the same general aerodynamic and physical rules. In later chapters, we will see that the physical properties required by wings are a major constraint on the evolution of animals' wings.

Scientists who study flight and engineers who design airplanes treat wings as if all wings are very complex, specialized structures. Yes, effective and efficient wings are rather specialized, but any more-or-less flat surface can act like a wing. Take a human hand, for example. When I am a passenger in a car with the windows down, I sometimes "fly" my hand out the window. I hold my hand flat, fingers together, palm down, in the onrushing air. If I tilt the front of my hand up, my hand is pulled up. If I tilt the front of my hand down, my hand drops. My hand is acting as a wing; not a very efficient one but one that generates the same basic categories of force as an eagle's wing or a bumblebee's wing. Tilting the front edge up represents what an engineer would call "increasing the angle of attack," which is one of the ways that lift is increased on any wing. My hand feels more lift and is pulled up.

## PHYSICS OF WINGS

Lift is one of the two important forces on a wing. In aerodynamics, lift is a force at right angles (perpendicular) to the direction of motion. So a wing moving horizontally would experience lift directed vertically upward. (If a wing moves at some angle other than horizontal, the lift will be tilted away from the vertical, which we will see is critical to the function of flapping and thrust production.) The other important force is the intuitively named "drag." Drag is any force that tends to slow the object as it moves through the air; it is a parallel force in the opposite direction from the motion. In this chapter, we will see that drag can be considered the "cost" of producing lift.

**Producing Lift**

Perhaps you have read or been taught that wings are curved on top and flat on the bottom, forcing air to move faster over the top; when that happens, Bernoulli's equation tells us that the faster air causes low pressure on top. This low pressure zone produces an upward force that we call lift. This traditional explanation is partly true, but it is incomplete and rather misleading. If it were true, wings would not work upside down (which would make most air shows rather boring), and several types of perfectly functional wings that are not flat on the bottom would not work. Yes, most wings are more convex on top than on the bottom, but this shape is not essential for lift production.

Rather than going into Bernoulli's equation or other technical aspects of aerodynamics, we will take a more intuitive approach. Imagine a flat plate such as, say, a sheet of plywood. If I lay the plywood sheet on a small cart so that it is parallel to the ground and I move it horizontally through the air, nothing much happens. But if I prop up the plywood so the front or "leading" edge is raised up a bit and then push it through the air, the plywood feels an upward force that we call "lift" (Fig. 3.1). The strength of the lift depends on how much I prop up the plywood: the lift gets stronger the higher I tilt the leading edge, at least up to about 15 degrees. If I raise the leading edge beyond that, lift suddenly gets much weaker. Finally (perhaps this is obvious—or perhaps not), the force on the plywood is the same whether I push the cart through the air or blow air over a stationary cart. When we try to visualize how wings operate, imagining the wing standing still with air blowing over it is often easier than imagining the air flowing around the wing as it moves through the air. Just as in sailing, the "relative wind" is the wind the plywood feels as it moves through the air, and in flapping, the wing's orientation to the relative wind can be crucial.

When I tilt the leading edge of the plywood sheet up, the plywood deflects air downward. Newton's third law says that for every action there is

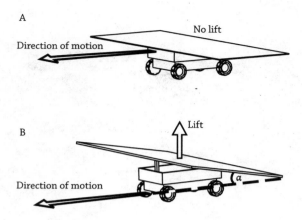

**Figure 3.1:**
A. A flat surface moving through the air without any inclination does not experience lift. B. If the same surface is tilted up by some angle α (the "angle of attack") and moved through the air, it will now experience an upward force, lift. (The angle of attack has been exaggerated slightly for clarity.) (Courtesy of Kevin Alexander, used by permission.)

an equal and opposite reaction, so the air the plywood sheet deflects down exerts an upward force on the sheet. This upward force is called "lift," and it is at right angles (perpendicular) to the movement of the plywood. Intuitively, we might think that the air is deflected downward by hitting the bottom of the plywood sheet and bouncing downward, like billiard balls hitting the side cushion of a billiard table; but this is not what happens. In fact, for any surface to act like a wing, the air flowing off the top and bottom surface must flow smoothly off the back or "trailing" edge (and for this to work properly, the flow over the top matters more than the air "hitting" the bottom). If the orientation of the trailing edge causes the air to flow downward as well as backward off the trailing edge, that downward component leads to lift production.

Remember, I said earlier that the lift increases as I tilt up the plywood (as I increase the "angle of attack," Fig. 3.1), but only up to about 15 degrees. If I increase the plywood's angle above some critical angle, the air stops flowing smoothly over the upper surface and peels away. Rather than following the upper surface, the air flows straight back, leaving a large turbulent region right above the upper surface (Fig. 3.2). This phenomenon is rather confusingly called "stall" even though it has nothing to do with engines. A stalled wing feels a sharp loss of lift and a simultaneous increase in drag (a slowing force); the turbulent air on top of the surface produces a broad, turbulent wake that actually causes a slight suction pulling back.

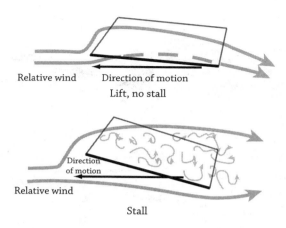

**Figure 3.2:**
At low angles of attack, the air flows smoothly over the surface (top), but if the angle of attack becomes too large, air flow over the top is disrupted, and lift drops off sharply (bottom).

The actual stall angle for any wing—plywood plate, bumblebee wing, jet airliner wing—depends on many factors, including size, speed, and shape. Larger, faster wings tend to stall at lower angles. The airliner's wing probably stalls at about 13 degrees, and wings of large birds like eagles or swans might stall in the 15-degree to 20-degree range. Small, slow wings, in contrast, stall at much higher angles. A bumblebee wing may reach 35 degrees or 40 degrees before stalling, and a fruit flight may not stall till its angle of attack exceeds 50 degrees.[1,2]

Although our plywood sheet acts like a wing in many ways, it is not very efficient: it does not produce nearly as much lift as—and it produces more drag than—a well-designed wing of a similar size. So what needs to change to allow a wing to produce more lift and less drag? One useful feature is streamlining. Streamlining greatly reduces drag (except at very small size scales). A streamlined shape looks something like a stretched-out teardrop: blunt and rounded in front, widest about one-third of the way along its length (Fig. 3.3A). If we could slice an airplane or bird wing from front to back and look at the cut edge, we would see that this cross section, the "airfoil" shape, is clearly streamlined (Fig. 3.3). Depending on size and speed, streamlining can cut the drag by up to a factor of 10![2]

Some wings have a symmetrical, stretched-teardrop cross section (Fig. 3.3A), but most have airfoils with camber. A cambered airfoil is a bit more convex on top and a bit less convex (Fig. 3.3B) or even concave (Fig. 3.3C) on the bottom. A wing with no camber, such as the symmetrical

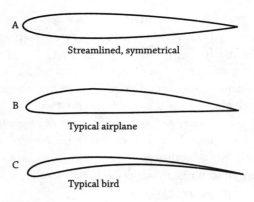

**Figure 3.3:**
Cross sectional shapes of three wings to show streamlining, with front or leading edge to the left.

airfoil of Figure 3.3A, does not produce any lift when it is directly aligned with the oncoming airflow (in other words, when the angle of attack equals zero); it must be tilted up to some angle above zero to generate lift. A cambered wing, however, produces lift even when directly aligned with the airflow—at an angle of attack of zero—and tends to produce more lift at any given angle than a wing with no camber. Although a few specialized airplanes have uncambered wings, most airplanes and all animals other than very small insects have wings with some camber.[3]

If a little camber is good, is a lot of camber better? Not necessarily. Imagine grabbing the leading and trailing edges of a wing with a symmetrical airfoil and bending them down (so the middle is forced up). That adds camber to the wing, and a typical wing might have "10%" camber, that is, the middle might be bent up above the no-camber shape by about 10% of the distance from leading to trailing edge. Too much camber increases drag and may even cause the wing to stall at low angles of attack. Fifteen percent camber is a lot for a wing (too much for many); a little camber goes a long way.

### Adjusting the Amount of Lift

Several things affect how much lift and drag a wing produces. Increasing angle of attack and camber (up to a point) increases lift, albeit at the cost of increasing drag. Increasing speed also increases lift. In fact, lift is proportional to the square of the speed, which means that doubling the speed would cause a fourfold increase in lift. So one way for a given wing to lift a heavier load is to fly faster, and a modest increase in speed can produce a healthy increase in lift.

One final wing property that affects lift production is wing area. This makes sense: a larger wing should provide more lift, and it does. Whereas the area of an airplane wing is essentially constant, birds are masters of adjusting wing area to change the amount of lift for a given situation. For instance, to descend rapidly, a bird can partly or completely fold its wings and drop sharply. Birds' ability to change wing area, as well as speed and angle of attack, gives birds a degree of maneuverability far beyond anything possible for airplanes.

**Drag and the Lift-to-Drag Ratio**

The "cost" of producing lift on a wing is an increase in drag, the slowing force. Even when a wing produces little or no lift—say, an uncambered wing moving at an angle of attack of zero—it still experiences drag, just as I experience drag when riding my bicycle fast. Lift production, however, adds on extra drag, and as lift increases, this extra, so-called induced drag also increases.[2] Induced drag thus represents the main cost of generating lift. The amazing thing about wings is that even with induced drag, wings can produce a lot more lift than drag. Under typical flight conditions, a run-of-the-mill wing might produce 10 times more lift than drag. To put this in concrete terms, if I push on this wing with 20 pounds of force to overcome its drag, it rewards me with 200 pounds of lift. In fact, the "lift-to-drag ratio," or what scientists call "L/D," is more important than the amount of lift itself produced by a wing.[1] A wing may produce huge amounts of lift, but if that wing requires overcoming huge amounts of drag, then it is not a very efficient or effective wing. The lift-to-drag ratio is thus a key measure of the effectiveness of a wing.* For most purposes, a high lift-to-drag ratio is a Good Thing. What kinds of wings have high lift-to-drag ratios? Induced drag is generated by aerodynamic processes that converge at the wingtip, so a shape that somehow minimizes the amount of wingtip should reduce induced drag. In fact, long, narrow wings tend to have higher lift-to-drag ratios than shorter, broader wings. All else being equal, a long, skinny wing will have a lower

---

* The L/D is actually not constant on a wing; it changes with angle of attack. What scientists and engineers actually use to characterize a particular wing is the wing's *maximum* L/D.

induced drag—and so a higher lift-to-drag ratio—than a short, broad wing of the same area. Scientists use "aspect ratio" to indicate the narrowness of a wing, where the aspect ratio is the wingspan—distance from tip to tip—divided by the average wing chord, or distance from front to back of the wing. Long, narrow wings are called high aspect ratio wings. They effectively have less tip for the amount of wing area than stubby, low aspect ratio wings, so high aspect ratio wings have less induced drag.[4] Of course, any wing—bird, bug, or airplane—is the result of design tradeoffs. When other factors—structural strength, say, or ease of folding—override sheer aerodynamic performance, a low aspect ratio wing might be the best overall choice. Scientists can infer a surprising amount about the biology of extinct flyers like pterosaurs from the aspect ratios of their wings (Chapter 8).

## Scale Effects

Humans are near the large end of the animal-size scale; yet even at a sprint we are not very fast compared to a running deer or a flying robin, so we do not experience noticeable air drag. In contrast, drag—and particularly the way it changes with size—can be of paramount importance for small flying animals like insects. As animals get smaller and move slower, the air feels more viscous. The air is no different for me or a mosquito, but I am so big that pressure-related processes swamp viscosity-related processes, and I hardly notice air's viscosity. For the mosquito, in contrast, forces that depend on pressures decline in importance so the viscosity becomes quite apparent.* (See Box 3.1. The "Reynolds Number.") This effect of size also affects wing performance. The main consequence is that for tiny flyers, streamlining is less effective and wings have much lower lift-to-drag ratios. In a nutshell, at small scales, wings are simply less effective. For example, a turkey vulture might have a lift-to-drag ratio of 15:1 (on the high side for birds); a dragonfly's lift-to-drag ratio probably would not exceed 5:1, which, even though quite high for an insect, barely makes it into the low end of the bird range.[5]

---

* Viscosity is actually a form of friction and should not be confused with density: honey is hundreds of times more *viscous* than water in spite of being just slightly denser whereas mercury's viscosity is not too different from the viscosity of water, but it is about 14 times *denser.*

*Box 3.1:* THE "REYNOLDS NUMBER"

A person can't study animal aerodynamics or hydrodynamics very long without running into the concept of the "Reynolds number." The Reynolds number, given the symbol *Re*, is a dimensionless index that indicates whether pressure drag or viscous drag is dominant. Viscous drag comes from viscosity. Viscosity is the resistance of air (or any gas or liquid) to flowing past itself or a solid surface, a sort of three-dimensional friction. Pressure drag is dominant at high Reynolds numbers, and viscous drag is dominant at low Reynolds numbers. Streamlining can dramatically reduce drag at high Reynolds numbers, but at low Reynolds numbers, reducing surface area (being spherical) cuts drag more effectively. The Reynolds number equals air density times speed times body (or wing) length divided by viscosity; as size and speed (or both) decrease, the Reynolds number decreases, and vice versa. For some technical uses, the exact value of the Reynolds number matters, but most of the time the general order of magnitude (powers of 10) of the Reynolds number is all that counts. Insect wings have Reynolds numbers of tens to thousands; bird wings, tens of thousands to hundreds of thousands; and airplanes, millions to hundreds of millions. Airplane designers can get away with ignoring viscous drag, but animal aerodynamicists usually cannot.

Osborne Reynolds, the man who gave the Reynolds number its name, was not actually studying drag directly when he developed the original concept. He was looking at water flows in pipes. He discovered that with slow flows or narrow pipes, water moved smoothly through the pipes, but with fast flows and larger pipes, the flow became turbulent and lost its smooth, laminar character. Reynolds proposed what became the Reynolds number as a way to predict whether a given flow would be laminar (smooth) or turbulent (full of random eddies). These flows turned out to correspond to viscous-dominated versus pressure-dominated conditions, respectively, and so the Reynolds number applies to any kinds of flows, not just in pipes.

Nowadays, the Reynolds number is most often used as an aerodynamic scale factor. For example, two wings with the same Reynolds number will have the same air flow patterns and the same ratios of forces. Conversely, two geometrically similar wings with very different Reynolds numbers will experience very different airflows, and force ratios such as the lift-to-drag ratio will be very different. At low Reynolds numbers, the flow patterns that produce lift are weaker, so lift and lift-to-drag ratios are lower at very low Reynolds numbers. In contrast, wings tend to be more effective and efficient at high Reynolds numbers.

## GLIDING

Gliding is unpowered flight. When a heron or a dragonfly drifts along through the air without flapping its wings, it is gliding. Gliding sometimes gets short shrift in animal flight discussions, but gliding was probably a stage in the evolution of flight for some animals, so we need to know something about gliding to look at the evolution of powered flight. Indeed, whether birds went through a gliding stage or not has been a source of much argument for decades, and at least some of the controversy may be related to imperfect understanding of flight mechanics, including gliding.

### Gliding Basics

Gliding, to continue my earlier analogy, shares many similarities with coasting downhill on a bicycle. A bicycle can coast downhill continuously, but if the road levels out or turns uphill, the bike slows to a stop unless the rider begins to pedal. Gliding works the same way: as long as a glider descends through the air, it can keep flying; but if the glider levels out or tries to climb, it will slow down and eventually stall. Perhaps, rather than "unpowered," we should think of gliders as "powered by gravity." Just as gravity pulls a coasting bicycle down a ramp, gravity pulls a glider down as well. One difference between a glider and a coasting bike, however, is that the characteristics of the glider's wings determine how steep a ramp it glides down. Curiously, it turns out that a gliding wing's lift-to-drag ratio sets its glide angle. A bit of fiddling with the geometry shows that the lift-to-drag ratio must be the same as the glide ratio, or the distance the glider moves forward for every meter it descends.[5] So a glider with a lift-to-drag ratio of 10:1 also has a glide ratio of 10:1, meaning it moves forward 10 meters (33 feet) for every 1 meter (3.3 feet) it descends.*

One peculiar and surprising consequence of the glider's lift-to-drag ratio setting its glide angle is the fact that changing a glider's weight does not have much effect on its glide path. Changing a glider's weight (without changing anything else) changes its *speed* but not its *glide angle*. If I don't change the glider's geometry in any way, I can add lead weights and it will fly faster but land just as far away as before I added weights. In other words, it will glide just as far, but it will get there faster! This effect

---

* The mathematically inclined reader will no doubt see that the glider's angle of descent or "glide angle" can easily be calculated using trigonometry; for a glide ratio of 10:1, the cotangent of the glide angle is 10, giving a glide angle of 5.7 degrees.

actually has more general consequences: as a rule of thumb, heavier flyers (whether gliding or powered) tend to fly faster than similar but lighter flyers.

If a glider's goal is to glide long distances, then a high lift-to-drag ratio is clearly beneficial because it produces a shallow glide angle. Both a wing's shape and its size play a big role here. High aspect ratio wings have high lift-to-drag ratios, but even high aspect ratio wings have diminishing lift-to-drag ratios as they get smaller. Few insects glide, and the ones that do—dragonflies and damselflies, for example—are large insects and have wings with high aspect ratios. A number of animals have evolved to glide with wings of low aspect ratio, such as flying squirrels. These are all much larger than insects, however, and their stubby wings come nowhere near the effectiveness of bird or bat wings (a colleague calls them "glide-assisted jumpers"). In short, if you are big, then even stubby wings may be effective enough for some limited gliding, but if you are tiny, even long, skinny, high aspect ratio wings may not be efficient enough for extensive gliding.

### Soaring

"Hold on," you say, "I've seen buzzards and hawks gliding on motionless wings, and they can go *up*. I thought you said gliders have to go down." These birds really are gliding, but it is a special form of gliding called "soaring." The birds are still going down *relative to the air*, but they are taking advantage of rising air. If the air rises faster than the bird descends, the bird can actually ascend while gliding. Returning to our coasting bicycle analogy, if, instead of coasting down a ramp, our bicycle coasts down an inclined conveyer belt, imagine what happens if the conveyer belt runs uphill. When the conveyer belt runs uphill faster than the bicycle coasts down, the bike will actually be carried up the belt. This is exactly what a soaring bird does: it seeks rising air—warm air heated by the ground or wind blowing up a steep slope—and climbs by riding the rising air upward.

### Wingless Gliders?

About a decade ago, then-graduate student Stephen Yanoviak was studying ants by observing them from catwalks built high in the rain forest canopy in Peru. He made the rather startling discovery that if he knocked an ant off a branch, the ant somehow steered its fall toward the nearest tree trunk, often landing on the trunk. These were worker ants with no trace of wings, so their ability to "fall" back to the trunk was astonishing. By coincidence, one of the faculty members in Yanoviak's department was Robert Dudley, an animal flight specialist who just happened to do much of his research in the tropics. Dudley was amazed by Yanoviak's story, and together they designed a series of experiments to see if ants really could guide their fall back to tree trunks, and if so, how they did it.

In their original experiments, 85% of the ants landed on the trunks of trees they fell from. Several studies later, Yanoviak and his colleagues have established that in addition to the original ant species they studied, a variety of other animals living in rain forest canopies have this ability. These include the workers of several unrelated species of arboreal ants from Africa and Central and South America as well as a variety of other wingless insects and spiders.[6-8] Curiously, the first species of ant they studied steered its falls backward (tail first), and seems mainly to use its hind legs for steering;[9] almost all the species they looked at later steered reassuringly head first.

Aerodynamically, worker ants are lousy gliders, with a glide angle of 75 degrees from the horizontal, compared with winged gliders, with glide angles of 10 degrees or 20 degrees. In fact, traditionally, biologists have called glide angles of greater than 45 degrees "parachuting" and reserved "gliding" for angles of less than 45 degrees. This is an arbitrary distinction with no basis in aerodynamics because any angle less than 90 degrees requires some lift production, so we will consider these ants to be gliding, albeit at a very steep glide angle.

The researchers dubbed this ability "directed aerial descent" to refer to animals with no apparent aerodynamic specializations that nevertheless perform a very crude, inefficient, but useful form of gliding. They suggest two important implications for their discovery. First, given the variety of wingless animals that display this ability, the behavior must have significant survival benefits. Indeed, small animals adapted to life in the tops of rain forest canopy trees are poorly adapted to life on the ground and tend to be easy prey for terrestrial predators. Moreover, worker ants generally do not survive unless they can return to their own colony's nest, so landing back on the trunk of their original tree is at a premium.

Second, if wingless animals with no noticeable aerodynamic adaptations can consistently and reliably adjust the direction of their descent during a fall, then they already possess most of the sensory and neural mechanisms they would need to control a more efficient glide. No one would think of a cat as a glider, yet cats reliably land on their feet after even short falls. That kind of orienting-during-a-fall behavior is directed aerial descent at its most rudimentary, and many arboreal animals have taken it a step further and evolved the ability to steer during falls. This ability could be a key exaptation* in the evolution of gliding. If many arboreal animals have this ability, as now seems likely, and if some of those animals experienced selection pressure to extend falls into glides, they would have a head start in evolving more effective gliding. Ironically, biologists have long considered the evolution of flight control ability to be one of the major hurdles to be overcome during the evolution of flight,[10] but this "hurdle" may already be behind many arboreal animals.

## APPLYING POWER

Flyers need to fly under power to take full advantage of flight and to overcome the limitations of gliding. Birds in powered flight flap their wings; if they are not flapping their wings, they are not powered and are gliding. Airplane wings may flex a bit in heavy turbulence, but airplane wings certainly do not flap the way birds flap their wings (at least, not more than once!). So airplanes can fly perfectly well without flapping their wings. Why, then, do flying animals need to flap their wings in order to fly under power?

### The Function of Flapping

Flapping is all about *thrust* production. "Thrust" is the force that moves a flyer forward by overcoming the force of drag on the wings and body. An airplane uses its wings to produce lift but it has separate devices—engines, jet, or propeller—to generate thrust. Birds do not have separate engines, so they have to use their wings to produce both the upward force and the forward force. Simply moving a wing through the air causes it to produce lift (that's how gliding works), so logically, flapping must be entirely for producing thrust.

In powered flight, the wings of flying animals do not work all that much like airplane wings. A bird's or an insect's wings actually operate much more like a helicopter's rotor blades. Rotor blades are long, narrow, highly specialized wings, and the helicopter's rotor generates both lift and thrust, much like animal wings. Moreover (and unlike airplanes), helicopters can

---

* Recall from Chapter 2 that an exaptation is something evolved for one function that can also perform or be co-opted for a new, second function. Biologists originally called these "preadaptations," but most have abandoned that term due to its teleological overtones.

use their rotors to hover, just as many flying animals can use their wings to hover. Helicopter rotor blades are not an exact model for animal wings, however: rotor blades rotate continuously, whereas a flapping bird wing cannot rotate continuously in one direction, so it periodically reverses direction in a cycle of upstrokes and downstrokes. Flapping wings operate differently during the upstroke and the downstroke parts of the cycle.

Actually, flying animals do not just move their wings up and down when flapping. During the downstroke, the wing moves down and forward with the leading edge tilted slightly down. On the upstroke, the leading edge of the wing tilts up and the wing moves up and back relative to the animal's body (Fig. 3.4). The wingtip thus follows an inclined path, from up and back at the top of the upstroke to down and forward at the bottom of the downstroke (Fig. 3.4). In general, almost all of the lift (including the upward force) and some or all of the thrust are produced during the downstroke. Larger flyers—herons, geese—may have a "passive" upstroke where they produce little or no useful force. Smaller flyers—hummingbirds, honeybees—may

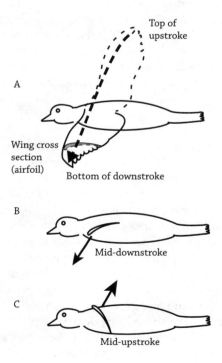

**Figure 3.4:**
Wing movements during flapping. A. Wing moves down and forward during the downstroke. B. Wing is inclined downward during the downstroke. C. Wing is inclined upward during the upstroke. (Solid arrows show direction of wing movement relative to body.)

have an "active" upstroke that produces a significant amount of thrust.[5] (See Figure 3.5 and Box 3.2. Flapping and Forces.)

**Power and Speed**

Powered flight has a major built-in difference from terrestrial locomotion in how speed affects power requirements. Whether I am on foot or driving

---

*Box 3.2:* **FLAPPING AND FORCES**

DOWNSTROKE

As the wing moves down and forward, it is tilted so that the lift is tilted forward (Fig. 3.5A). The lift, remember, is perpendicular to the direction of movement of the wing, so it is no longer straight up on a flapping wing. This forward tilt of the lift is where the thrust comes from. We can separate the lift into its upward component (the upward force balancing the animal's weight) and a forward component, the thrust. (I am ignoring the wing's drag for simplicity, but the drag is so much smaller than the lift that it does not significantly change the general pattern.) Most of the useful flight forces—the upward force and the thrust—are produced during the downstroke, so it tends to last noticeably longer than the upstroke.

UPSTROKE

Figure 3.5B shows the passive upstroke, typical of a large bird. This upstroke produces little useful force because the wing is aligned directly with the air flow. A passive upstroke is mainly a return stroke: it is just the movement necessary to move the wing into position for the next downstroke with as little effort as possible. It may or may not produce a bit of upward force but it does not contribute any thrust.

Active upstrokes vary in how much useful force they generate, but the extreme is shown in Figure 3.5C. If an animal can tilt its leading edge straight up and move its wing up and back very rapidly, it can actually produce a significant amount of thrust. This process is only really effective if the flyer can move its wings backward faster than its body flies forward through the air (see wingtip path, Fig. 3.5C). Insects, hummingbirds, and very small bats are probably the only flyers that can take full advantage of active upstrokes in normal flight—although some larger flyers may be able to perform such a stroke when flying slowly—or take advantage of less extreme (and less effective) variations.

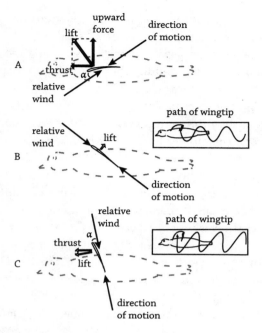

**Figure 3.5:**
Forces on a flapping wing. A. Downstroke; lift has upward and forward (thrust) components. B. Passive upstroke, used by medium-sized and large birds in fast forward flight; forces minimized. C. Active upstroke, used by smaller flyers, especially in slow flight; negative angle of attack generates forward-tilted lift on bottom of wing. Insets: path of the wingtip through the air.

a car, if I want to go faster, I need to put more power into moving. Going slower requires less power, and stopping requires essentially no power. (Yes, a bit of power goes into keeping the engine running in my car or keeping my body alive even at rest, but this is a drop in the bucket compared to the power needed to move.) In a nutshell, standing still uses no power, and moving requires more power the faster I go.

Flight has fundamentally different power requirements. At higher speeds, as I fly faster I need more power, just like running. Strangely, as I slow down, the power requirement at first drops, but then as I fly slower and slower, I begin to require more and more power. If I am in an airplane and I slow down below a certain speed, I actually have to increase the throttle to fly level at slower speeds.* If I am a bird, I have to flap my wings faster and harder as I slow down below a certain speed, if I want to fly level.

---

\* Test pilots call this "being on the back side of the power curve," and it is a very bad place to be in most airplanes.

To us pedestrian creatures, the peculiar relationship between flight power and speed is quite foreign. Scientists call it a "U-shaped power curve" because a graph of power against speed is high at both low and high speeds and low in the middle, thus making a U-shape.[11] This relationship has several practical consequences for flyers. One is that any flyer has some specific intermediate speed that requires the least effort to fly. For example, the minimum power speed for a swallow is about 3½ meters per second (just under 8 miles per hour), and for a kestrel (small falcon) it is about 5 meters per second (11 miles per hour). These minimum power speeds—flight speeds that require the least effort—are as fast as or faster than the *maximum* running speeds of similarly sized terrestrial animals. Curiously, the speed of minimum power (least effort) is not the speed that gives best "fuel economy," that is, that uses the least food energy per distance traveled. The best fuel economy speed is the speed that allows the flyer to fly the greatest distance, and the speed for best fuel economy is a bit higher than the speed of minimum power. Although this sounds contradictory, at speeds a bit higher than the minimum power speed, the reduction in travel time more than compensates for the increased energy consumption. Scientists have observed birds in the middle of long-distance migrations on radar, and these migrants mostly fly at speeds quite close to their estimated best-economy speeds, even though the flyers would feel that they are exerting more effort than flying at the speed of minimum power.

If flying slower and slower requires more power, what about coming to a complete stop in flight? In other words, what are the power requirements of hovering? As you probably guessed, hovering requires a *lot* of power. Hovering may be the most power-hungry activity in the animal kingdom; it is certainly the most power-hungry activity in animal (and mechanical) locomotion. The main reason that hovering requires so much power is that the wings get no air moving over them due to the body's movement through the air; the animal's body is not moving. Instead, the animal must use its wing muscles to move the wings fast enough for them to produce enough lift. In other words, in forward flight, a bird needs only enough power to overcome its drag so the wings can move through the air, but when hovering, the bird needs enough power to move the wings very rapidly to make up for the lack of air flow from forward flight.

Hovering requires some rather contradictory structural properties. Think about how an animal's weight is suspended in flight: the animal's body literally hangs from its wing joints. A flying bird's weight is carried entirely by its shoulder joints, so these joints must be strong, robust structures. A loose, highly flexible joint may not stand up to the loads of flight because its flexibility limits how rigidly it can be supported. Yet hovering requires a highly flexible wing joint so that the wings can make the necessary, exaggerated flapping movements—hovering normally requires both a larger wing stroke and a lot more wing twisting than forward flight. So in order to hover, the hoverer must have a strong enough wing joint to carry its whole body weight, but the joint must also be flexible enough to allow the greater degree of movement that hovering requires. As we will see shortly, both power and structural constraints make hovering easier if you are small. Insects and hummingbirds are masters of hovering and can do so for long periods. Sparrows and finches can hover for a few seconds, but geese and eagles are simply too large to hover.

One subtle consequence of the speed-power relationship of flying is that the minimum-power flight speed is so high that it can make flight very economical, much more so than walking or running. Even though flight requires lots of power, flyers travel so much faster that they actually use less total energy than a runner. A sparrow may use 5 times more power to fly than a mouse needs to trot, but if the sparrow flies 10 times as fast as the mouse trots, the sparrow will consume half as much energy (fuel) as the mouse to go the same distance. In other words, a flyer typically consumes much less energy to go a given distance than a similarly sized runner.

## GLIDING VERSUS FLAPPING: FALSE DICHOTOMY?

So far I have described flapping flight and gliding flight as two completely separate activities. This division seems reasonable given the way I have defined the two modes of flight; it has even been used as an argument against gliding as a possible precursor of flapping. Some scientists have argued that weak or poorly developed flapping would be so ineffective that it would provide no benefit to gliding animals (or might even be less effective than gliding), so gliding could not have led directly to flapping.[12, 13]

In fact, both theoretical modeling and experiments using flapping robots show that even low-amplitude, weak flapping can produce enough thrust to be useful, even when such flapping is too weak to maintain level flight.[14-16] These results mean that we must think of gliding versus fully powered flight—flapping flight as used by living birds, bats, and insects—as two extremes on a continuum. Between these extremes, animals could use flapping with a range of effectiveness, from weak flapping to slightly extend a glide to stronger flapping that might increase glide distance by five- or tenfold. Although living animals appear to use either pure gliding or fully powered flapping flight, the evidence suggests that at least some (possibly all) of the living flapping flyers went through a partially powered stage. For clarity, from this point on, when I use the terms "flapping flight" or "powered flight," I am referring to the fully powered extreme of the continuum, and I will use "weak flapping" or "partially powered flight" to refer to intermediate behavior between gliding and fully powered flight.

## SIZE MATTERS

Just as size affects the hovering ability of flyers, size affects other facets of flight in important ways. Size will figure prominently as we look at the evolution of flight. Nature's flyers range from insects with wingspans of little more than a millimeter (1/10 inch or so) to the largest animals that have ever flown, such as pterosaurs and the vulture-like teratorns, with wingspans in the 8- to 12-meter (25- to 40-foot) range. The differences between being really small and being really big matter a lot more for flyers than for walkers. The physics of a cockroach's legs are not all that different from the physics of a horse's legs. Flying is different. Although their wings work basically the same way, small flyers face a rather different suite of constraints and opportunities than do large flyers. Bees can do things that buzzards can't, and vice versa.[2]

Hovering, for example, strongly favors small flyers for at least a couple of reasons. For example, the surface-to-volume ratio greatly favors small animals, as we saw in Chapter 1. As animals get smaller, their weight goes down much faster than their surface area. For animals of the same general shape, cutting the body length in half will reduce the wing area to one-fourth but the weight to one-eighth of the original values. Smaller flyers are effectively supporting lighter bodies with larger, lighter wings, so they don't have to move those wings nearly as fast in order to hover. Insects take this a step further: with their unusual, corrugated structure, insect wings are remarkably light even for their small size while still being exceptionally stiff and strong. In addition to overall geometry, changes in muscle cross sectional area (Chapter 1) and muscle mass with changing body size may play a role in hovering ability as well. In concrete terms, insects and hummingbirds can hover easily, robins and pigeons may be able to hover briefly, but crows probably cannot hover at all.

Large animals suffer a double whammy with hovering, because they are also at the wrong end of the structural scale. Big wings are disproportionately heavy: double their length and their surface area will increase by four times (good) but their weight will increase by eight times (very bad). (See Box 3.3. Scaling of Size.) Heavy wings have two drawbacks for

> *Box 3.3:* SCALING OF BODY SIZE
>
> If animals retain the same basic shape but change in size, different measures of size change at different rates. To take the simplest shape, basic geometry says that if I double the diameter of a sphere, the sphere's surface area increases by a factor of 2-squared or 4 and its volume increases by a factor of 2-cubed or 8. The same would apply to two animals of the same shape if one were twice as long as the other. An animal's weight is normally directly proportional to its volume, so if an animal's length doubles, its weight goes up by eight times. If one animal were three times longer than another of the same shape, the big animal would weigh 3-cubed or 27 times more than the small one.
>
> Flying animals span a stunning size range, from tiny insects like thrips and parasitic wasps with wingspans of less than 1 millimeter (1/25 inch) to giant pterosaurs with wingspans of over 11 meters (more than 35 feet). This is a 10,000-fold range on the length scale; if thrips and pterosaurs were the same shape, they would differ in weight by a factor of about one trillion-fold. Because pterosaurs and thrips are not the same shape (and also because wingspan, as opposed to body length, may exaggerate linear dimensions) pterosaurs are "only" about a billion (1,000,000,000) times heavier than thrips. So we cannot predict exactly how much disproportionately heavier flyers get with increasing size; clearly, both surface area and weight increase *much* faster than linear dimensions as animals get bigger.
>
> Because of the way weight increases as flying animals become larger, big flyers face conflicting priorities: they need to be as light as possible to make flight easier, but they need disproportionately stronger wing structures to carry their disproportionately heavier body weight. The structural requirement leads to wings becoming even *more* disproportionately heavy. Very heavy flyers thus tend to have "overbuilt," heavy wings compared to their smaller relatives; by the same token, very small flyers tend to have amazingly simple, lightweight wing structures. This pattern shows up even within insects: a large wasp like a hornet will have four or five main supporting wing veins (providing the main structural support for the wing), whereas a tiny parasitic wasp will only have one or two wing veins.

hovering. First, they require a stronger, stouter joint, which works against the flexibility required by hovering. Second, heavy wings have a lot of inertia, so the animal must work harder to move them. Hovering requires faster and larger wing movements than forward flight, so heavy wings hinder hovering. Hovering is clearly the province of small flyers.

Gliding, in contrast, is where big flyers come into their own. Because of the way viscosity affects flyers of different sizes, small wings produce more drag for a given amount of lift, or to turn it around, big wings produce less drag for a given amount of lift (see Box 3.1). All else being equal, a big wing—say, a wing with a 200-centimeter (79-inch) span like a stork—will have a higher lift-to-drag ratio than a small wing—say, one with a 6-centimeter (2⅖-inch) span, such as a dragonfly. Fruit flies have miserable lift-to-drag ratios of around 2:1, and larger insects probably do not get much higher than 4:1 or 5:1. (Remember, a lift-to-drag ratio of 2:1 means that if the fruit fly tried to glide, it would move forward only 2 meters for every 1 meter it descends.) Birds range from about 4:1 on the low end for small, stubby-winged birds like sparrows, to around 10:1 for small birds with high aspect ratio wings like swallows, and large birds with moderate aspect ratios like herons and hawks. Big birds specialized for gliding run even higher: turkey vultures (buzzards), for instance, at about 15:1, and the largest albatross species at 19:1.[5]

The most effective soaring animals must be good gliders, so the real masters of soaring, the ones that depend on soaring as much as or more than on flapping, are all big. Vultures, hawks, and eagles are probably the most familiar of these specialized soarers, which also include the biggest modern flying birds—condors (heaviest) and albatrosses (longest wings). These very large birds must get by with proportionately less powerful muscles—they are barely powerful enough for forward flapping flight—so they soar and avoid flapping as much as possible.

Knowing how size affects the physiology and aerodynamics of modern flyers allows us to understand the flight of ancient animals. A 300-million-year-old mayfly with a 3-centimeter (1-inch or so) wingspan could probably hover but certainly did not soar routinely (Chapter 5). In contrast, the 70-million-year-old pterosaur *Pteranodon*, with a 7-meter (23-foot) wingspan could not have hovered but probably spent most of its flight time soaring (Chapter 8). This "biomechanics" approach has, in recent decades, greatly expanded our understanding of how extinct animals lived, and has helped us understand what behaviors would have been probable, possible, or impossible for these ancient animals.

**REFERENCES**
1. J. J. Bertin and M. L. Smith (1979) *Aerodynamics for Engineers.*
2. S. Vogel (1994) *Life in Moving Fluids: The Physical Biology of Flow.*
3. D. E. Alexander (2009) *Why Don't Jumbo Jets Flap Their Wings? Flying Animals, Flying Machines, and How They Are Different.*

4. J. D. Anderson (2007) *Fundamentals of Aerodynamics*.
5. D. E. Alexander (2002) *Nature's Flyers: Birds, Insects, and the Biomechanics of Flight*.
6. S. P. Yanoviak and R. Dudley (2006) *Journal of Experimental Biology*.
7. S. P. Yanoviak, M. Kaspari, and R. Dudley (2009) *Biology Letters*.
8. S. P. Yanoviak, Y. Munk, and R. Dudley (2011) *Integrative and Comparative Biology*.
9. S. P. Yanoviak, R. Dudley, and M. Kaspari (2005) *Nature*.
10. J. Maynard Smith (1952) *Evolution*.
11. J. M. V. Rayner (1999) *Journal of Experimental Biology*.
12. G. R. Caple, R. T. Balda, and W. R. Willis (1983) *American Naturalist*.
13. P. Burgers and L. M. Chiappe (1999) *Nature*.
14. U. M. Norberg (1985) *American Naturalist*.
15. R. L. Nudds and G. J. Dyke (2009) *Evolution*.
16. K. Peterson, P. Birkmeyer, R. Dudley, et al. (2011) *Bioinspiration & Biomimetics*.

CHAPTER 4

# Gliding Animals

*Flight without Power*

A small, mottled-brown lizard scurries up the trunk and out on a limb of a very tall tree in a rain forest in Malaysia. As the lizard noses around on the limb searching for insects to eat, it is startled by a bird landing nearby. The lizard leaps off the branch away from the tree, in spite of being over 30 meters (100 feet) off the ground. As if by magic, a stubby wing sprouts from each side of the lizard's body. After a brief dive to gain airspeed, the lizard flattens its glide and steers toward another nearby tree. Having picked out a likely landing spot on a large limb, the lizard rears up into a short, steep climb to slow down just before contact. Judging speeds and distances with exquisite precision, the lizard alights gently and scurries off in search of dinner.

This lizard is a specimen of *Draco formosus*, one of the dozen or so species in the genus *Draco*, commonly called "flying lizards" or "flying dragons." (The genus name, "*Draco*," is Latin for "dragon.") All the lizards in this genus have wings formed by elastic skin stretched over a set of springy, elastic, retractable ribs along the sides of their bodies (Fig. 4.1). When folded up, the wing is barely noticeable and appears not to interfere with the lizard's running or climbing. When extended, however, these wings generate enough lift to give the lizard about a 2:1 glide ratio—it glides forward two feet for every one foot it descends in a steady-state glide. Moreover, it can flatten its glide considerably by gradually increasing its angle of attack and decelerating.

These lizards are gliders. They cannot flap their wings so they cannot sustain level flight. To say that these animals are not "true" flyers, however, is to seriously underestimate their abilities. A purely gliding animal

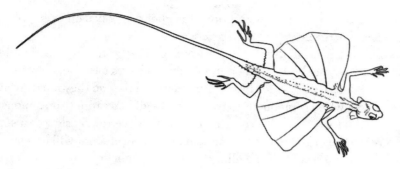

**Figure 4.1:**
*Draco* gliding, looking down on it from above. (Courtesy of S. T.)

needs to be able to do most of what a bird can do: steer, recover from gusts, avoid obstacles, detect safe landing spots, and perform soft landings. Unlike a bird, a glider's flights are always brief and always finish lower than they began.

A surprising diversity of animals (including even fish!) have evolved the ability to glide, and the extinct gliding kuehneosaurs, which looked much like modern *Draco*, were some of the earliest vertebrates to take to the skies. Moreover, gliding was most likely a step on the way to powered flight for at least some of the lineages of flapping flyers.

## GLIDERS VERSUS FLAPPERS

Gliding animals such as *Draco* have several features in common with flapping flyers like birds. Most obviously, both have wings. The wings of gliders and flappers are very different in structure, but both have large, flattened, cambered surfaces that produce lift using the same mechanism. Although not as mobile and flexible as a bird wing, the wings of almost all gliders are adjustable enough to use for steering, just the way a gliding bird or bat steers.

A more subtle similarity involves the "control system." Gliders need to be able to judge whether they can safely glide to a particular target, which requires both sharp vision and sophisticated mental processing. Once aloft, a glider needs to use that sharp vision along with quick reflexes to avoid objects and stay on course. The glider needs the brainpower to accurately control the glide and to determine just the right time for the pull-up to produce a soft landing. These are all basically the same as the sensory and mental tasks of a flapping flyer like a robin or a house fly, with the added necessity for the glider to always get it right the first time: if the robin arrives too low or too slow, it can always flap its wings to compensate, but *Draco* does not have that option. The glider thus has an even greater need for high precision and accuracy.

The fundamental difference between gliding animals and flapping flyers is that pure gliders, when flying, must always descend. In physical terms, gliders are propelled by gravity; they can't flap, so they must always have a downward component to their flight—gliders cannot continuously fly level or climb. Just as a coasting bicycle will eventually slow to a stop on a level road, a glider attempting to maintain level flight will decelerate and eventually stall. When a glider adjusts its flight so that all the forces (lift, weight, drag, thrust) are balanced, it will be in an equilibrium or non-accelerating glide. Typically, gliding animals in equilibrium glides have fairly steep glide angles, descending at approximately 30 degrees or more below the horizontal.[1] Gliding animals actually spend little of their time in equilibrium glides, however.

When *Draco* jumps off a tree limb, at first it falls steeply to accelerate to a high enough air speed for its wings to become effective. Scientists call this the "ballistic phase" of the flight. Once the lizard is going fast enough, it levels out, and then it begins to gradually increase its angle of attack and decrease its speed to maintain a flat glide, the "aerodynamic phase." In typical glides, *Draco* spends most or all of the aerodynamic phase decelerating, giving it a much flatter glide than when in an equilibrium glide (Fig. 4.2). When executed properly, the glider stalls just as it reaches its intended landing spot. For very long glides, or glides without a well-defined landing site, *Draco* may enter an equilibrium glide and just pull up into a brief, decelerating climb to slow down before alighting.

Many birds use rising air currents to stay aloft while gliding, staying up for hours and covering many miles in the process. This is called soaring (Chapter 3), and in principle, gliding animals could soar. Most gliders, however, have wings that appear almost square or circular from above, meaning they have very low aspect ratios. These wings are aerodynamically inefficient (they have low lift-to-drag ratio values), so gliding animals would need very high-speed updrafts—much higher than they would normally encounter and much higher than the updraft speeds used by soaring birds.

The difference between the aspect ratios (AR) of flapping flyers—such as birds and bats—and gliding animals—such as *Draco* or flying squirrels—are quite striking. Birds typically have aspect ratios of 6 or 8 (their wings are 6 or 8 times wider from tip to tip than from front to back). For instance, robins have aspect ratios of about 6 and mallard ducks have aspect ratios of about 8. Even birds with fairly short, broad wings, like sparrows (AR = 5.3), have considerably higher aspect ratios than gliding animals.[2]

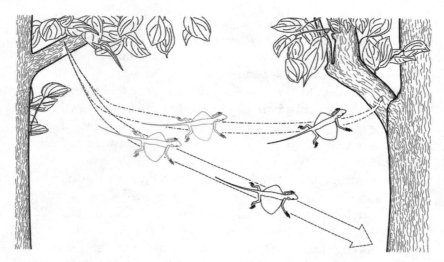

**Figure 4.2:**
Gliding trajectories. Width of the path arrow represents speed. (Courtesy of S. T.)

The aspect ratios of the wings of several species of *Draco* range from 1.7 to 2.3,[1] and flying squirrels have aspect ratios in the range of 1.2 to 2.2.[3] Thus, pure gliding animals generally have wings with aspect ratios that are too low for them to soar effectively.

Why do most gliders seem to be limited to wings of low aspect ratio? Although scientists have debated this question without reaching a clear consensus, the most reasonable explanation has to do with the body parts that make up the wing. In *Draco*, the trunk (torso) ribs have hinged extensions that support the wing. If these rib extensions were much longer, they would not have room to fold flat along the animal's flanks when not in use and would also get entangled with the hind legs. Flying squirrels, like all other mammalian gliders, form their wings from a stretchy fold of skin, the patagium, that runs from their front legs along their flanks to their hind legs. That arrangement limits their wingspan to the lengths of their outstretched legs. The legs of gliding squirrels are slightly longer than the legs of non-gliding squirrels, but not dramatically longer. Why haven't flying squirrels evolved much longer legs so they could have longer, narrower wings? Flying squirrels spend much more time climbing around on tree branches than gliding, and biologists suspect that if their legs became much longer, the length might interfere with climbing. Similar arguments apply to other gliders, such as frogs and geckos.

These low aspect ratio wings do have at least one virtue: they can operate at very high angles of attack without stalling. Functionally, this means that such a wing should be able to fly slower than a longer, narrower wing with the same area. In other words, as a glider like *Draco* slows down, it can keep increasing the angle of attack to maintain lift as it flies slower and slower, which should allow it to land at a lower speed than would be possible with a high aspect ratio wing. This feature is a distinct advantage for a flyer that cannot flap its wings in a braking action to slow down for landing.

In Chapter 3, I described wingless worker ants that can direct their falls so that they land on the trunk of the tree they fell from. Should we consider these ants to be gliders? They have no apparent wing-like aerodynamic surface and, at best, their "glide" is a steep 75 degrees or 80 degrees below the horizontal. Yet they do not fall vertically, and they can, without a doubt, actively steer their course. These ants use aerodynamic forces to direct and probably slow their fall. They may not be very efficient gliders, but what they do fits the aerodynamic definition of a glider with a very low lift-to-drag ratio (L/D) value. Moreover, ants aren't the only ones doing this. Other wingless arboreal arthropods—some spiders, bristletails, and immature insects—also have this ability.[4]

## WHO'S WHO? EXTINCT AND LIVING GLIDERS
### Extinct Gliders

Lizards and lizard-like reptiles have evolved a *Draco*-like gliding body form at least four different times. (Although some of them are not technically lizards, they are on the lineage leading to modern lizards and they had body proportions and postures generally similar to present-day lizards.) The earliest one we know about so far was *Coelurosauravus elivensis* from the late Permian period, about 252 million years ago (Fig. 4.3). Unlike the other lizard-like gliders, *Coelurosauravus* supported its wings with struts of dermal bone separate from its ribs, but externally its wing would have looked very much like *Draco*'s.[5] The late Triassic kuehneosaurids from about 220 million years ago include the largest—several species in the genus *Kuehneosaurus*—and the smallest—*Icarosaurus siefkeri*—of the lizard-like gliders.[6–8] The youngest (and most recently discovered) of these extinct gliding reptiles is *Xianglong zhaoi*, from the early Cretaceous period, about 127 million years ago.[9] Other than *Coelurosauravus*, all had wing membranes supported by modified ribs.

Intriguingly, all these fossil gliders appear to have had substantially longer wings for their size than *Draco*. Given that *Draco*'s wing length seems to be limited by the need for its wings to fold neatly along its flanks and not interfere with its hind legs, we are left to wonder exactly how these fossil gliders stowed their wings. Presumably these extinct animals received

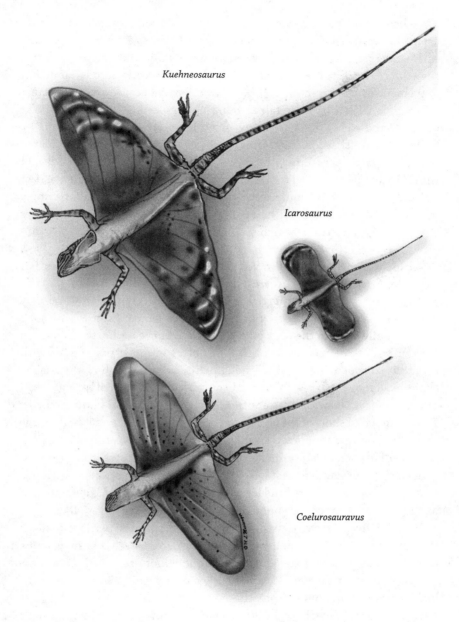

**Figure 4.3:**
Extinct gliding lizards and lizard-like reptiles, reconstructed to approximately the same scale; *Kuehneosaurus* had a wingspan of approximately 40 centimeters or 16 inches. Color patterns are entirely speculative but are inspired by patterns on wings of living species of *Draco*. (Courtesy of Lori Messenger, used by permission.)

some benefit from their long wings. These longer wings give them higher aspect ratios, which in turn give them higher L/Ds, so at first glance, these early gliders would seem to have more aerodynamically effective wings than *Draco*. In fact, the ancient gliders show very little aerodynamic improvement over their modern counterpart. Partly because all gliders start every glide with a ballistic fall to build up airspeed—where wings don't matter—and partly because the extinct gliders' wings were still short by bird standards, they appear not to have been especially efficient compared to *Draco*. Researchers Jimmy McGuire and Robert Dudley used models of the extinct gliders in wind tunnels to show that their slightly higher aspect ratios did not seem to confer any particular advantage other than simply increasing the wing area.[8] All else being equal, more wing area means lower wing loading, and the researchers found that lowering wing loading improved glider performance much more than increasing aspect ratio.

A couple of early reptiles evolved gliding with less *Draco*-like bodies. For example, *Mecistotrachelos apeoros*, of similar age to the kuehneosaurids, had a similar wing membrane to the kuehneosaurids but had a startlingly long neck—almost as long as its trunk or torso length.[10] Birds with long necks usually fly with the neck pulled into an S-curve to shorten it, but the neck of *Mecistotrachelos* was not that flexible, so it must have glided with its neck extended (Fig. 4.4). Although this posture might be unstable, slight head movements would have been very effective for steering.

*Sharovipteryx mirabilis* was another, even more unusual gliding reptile. This animal had short front legs and very long hind legs, and it had a large gliding membrane supported mainly by the hind legs. Scientists have puzzled over how *Sharovipteryx* held its legs in flight ever since it was first described by a Russian scientist back in 1971. The most recent suggestion is that *Sharovipteryx* held its hind legs out so that they formed a delta-shaped wing,[11] reminiscent of the supersonic Concorde airliner (Fig. 4.4)!

Very few fossils have been found of gliding mammals, perhaps because they tend to be small and delicately built and also because they may not be recognized as gliders if the wing membrane is not preserved. One spectacular exception is *Volaticotherium antiquum*, a petite, 13-centimeter (5-inch) primitive mammal from the Jurassic of China (the exact age is uncertain, but it is well over 130 million years old).[12] It appears to have been an arboreal animal well adapted to gliding between scattered trees in open woodlands. This animal is so ancient that it is not closely related to any modern mammals. The ancient age of *Volaticotherium*, plus the common occurrence of gliding among modern mammals, suggests that gliding evolves fairly readily in arboreal animals, so scientists suspect that a number of extinct gliding mammal species await discovery.

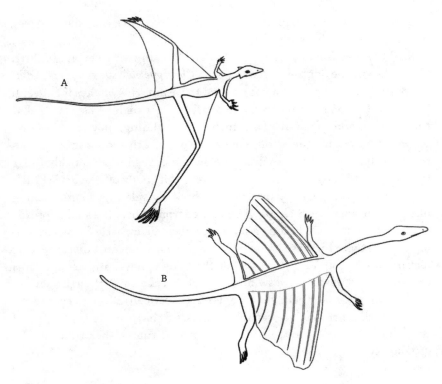

**Figure 4.4:**
Extinct gliders with strange anatomy. A. *Mecistotrachelos*. B. *Sharovipteryx*. (Courtesy of S. T.)

## Living Gliders

Given *Draco*'s rather small-looking wing, these creatures are surprisingly capable flyers. One biologist described a territorial male chasing an interloping male along tree branches and through the air, with the aerial part of the chase involving as much turning and twisting as the part along branches.[1] People have seen gliding *Draco* lizards perform sharp turns, U-turns, and even barrel rolls. Although their long tails undoubtedly help to stabilize their flight, these lizards seem to steer mainly using wing adjustments, just like a bird.

Several species of gecko also glide, and they also use a flap of skin alongside their flanks. In geckos, however, these skin flaps have no internal support, are not connected to the legs, and appear to be extended passively, entirely by air pressure.[13] They are much smaller than the wings of a similarly sized *Draco*, but geckos compensate at least partly with webbed feet. These webbed front and hind feet not only provide additional wing

GLIDING ANIMALS [67]

area, but geckos must also use them for steering, since they have no way to actively adjust the wing.

Snakes must seem like the least likely of all animals to evolve gliding, but snakes in the Southeast Asian genus *Chrysopelea* have done just that. Biologist John "Jake" Socha has studied these amazing snakes for several years and he was the first to try to analyze their flight.[14] He found that these snakes don't just fall off a branch to start gliding; they actually jump up, which may be unique among snakes. To prepare to jump, a snake allows the front half of its body to droop below the branch, then quickly flings the front of its body up, which lifts the rest of its body off the branch. The snake then forms its body into two or three big sideways loops, looking like a sort of squashed S-shape when viewed from above. The snake spreads out its ribs so its body becomes somewhat flattened, partly for streamlining and partly to increase its "wing" area. Perhaps the most disconcerting feature of *Chrysopelea* gliding is that the body is not static; it undulates rapidly, almost as if it were swimming through the air. Although they are not stellar gliders—their glide angles are between 20 degrees and 45 degrees—these snakes are capable of making sharp turns in flight, and they much prefer to land in other trees or shrubs rather than on the ground.

### Mammals

Gliding has evolved at least six times among living mammals: twice in rodents, three times in marsupials, and once in the colugos or "flying lemurs."

Among rodents, the 40 or so species of flying squirrels are extremely widespread, from North America to northern Europe and across central, south, and southeast Asia. They can be fairly common in mature forests, but most people rarely see them because many species are nocturnal.

The southeast Asian colugos are among the largest modern gliders and are also probably the most aerodynamically sophisticated. These ferret-sized, secretive gliders live in the canopies of some of the world's tallest rain forests in Malaysia, Indonesia, and the Philippines, so they have escaped attention from most scientists. Indeed, they are so poorly studied that they have long been known as "flying lemurs," even though they are certainly not lemurs and are not particularly closely related to lemurs.

Colugos glide on a large fold of skin that stretches from their front limbs to their hindlimbs and continues between the hindlimbs to include the tail; it also continues forward to the side of the neck (Fig. 4.5). Unlike other mammalian gliders, the colugo's aerodynamic surface includes the

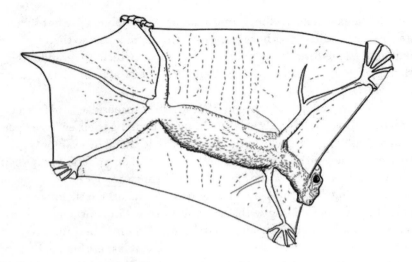

**Figure 4.5:**
Colugo gliding, showing the flight membrane (patagium) and webbed fingers. (Courtesy of S. T.)

hands and feet; they have highly elongated, webbed fingers and toes. Although we know very little about how colugos glide, for the hand to be incorporated into the flying surface hints at unusual maneuverability. Changing the shape of the tip of the wing can be a very powerful turn mechanism, and dexterous fingers would give colugos very fine, precise control of such a potent mechanism.

Colugos also have low wing loading for gliding animals (that is, large wings for their body weight), which allows them to glide more slowly for their size than other gliding mammals.[15] Adjustable wingtips and slow air speeds would make colugos extremely maneuverable. Thus, although we know very little about their behavior or ecology, colugos' anatomy suggests they are probably the most accomplished of gliders.

**Other Gliders**

At least two different lineages of frogs have evolved gliding. Gliding frogs may sound odd, but in many moist temperate and especially tropical regions, frogs live in trees. Moreover, frogs typically have webbed hands and feet, which make handy steering devices for directed aerial descent if they should happen to fall out of a tree. Their webbed feet are a nice example of an exaptation (Chapter 2): they originally evolved for swimming, but are quite useful for directed aerial descent with little modification. They thus provide the raw material for gliding adaptations involving greater surface area. Extant gliding frogs have elongated fingers and toes, so their webbed hands and feet form aerodynamic surfaces on which they glide. Such "flying frogs" are found in both South American and southeast Asian rain forests.

People once thought that oceanic flying fish (Exocoetidae) and freshwater hatchetfish (Gastropelecidae) could fly by flapping their enlarged pectoral fins as wings. We now know that exocoetid flying fish can only glide when airborne—although they do have the remarkable ability to dip their tails in the water while airborne and beat them to accelerate for another bout of gliding.[16] These oceanic flying fish are thus the only known animals that evolved gliding without living in trees. In contrast, after decades of anecdotes about tropical hatchetfish flapping their fins to fly in air, careful research showed that they are excellent leapers but get little or no benefit from their fins while traveling in air.[17] Not only do they fail to flap but they are not even any good at gliding.

Finally, as noted above, ants as well as several other kinds of wingless arthropods perform a rudimentary kind of gliding when they "skydive" back to the tree trunk after being dislodged or jumping from a branch.

## WHY GLIDE?

Given the enormous variety of animals that have evolved gliding ability, gliding must confer some advantage significant enough to outweigh the cost of growing the wing surface and carrying it around when not gliding. Probably the biggest advantage is escaping from predators. A glider can leap out of a tree and safely reach another tree trunk or branch that is much too far away for a non-gliding leaper to reach. So if a glider can detect a non-gliding predator in time, the glider will normally escape from such a predator with ease. This ability has an extremely strong evolutionary benefit, because being killed by a predator removes the prey individual's genes from the gene pool; an individual that can escape from a whole category of predators is much more likely to contribute genes to the next generation. Thus, many biologists suspect that escape from predators was the selective factor that initiated gliding adaptations in most gliding animals.[18]

Escaping from predators is not the only advantage of gliding, however. Arboreal animals commonly risk falling out of trees, and the risks of falling—and benefits of gliding—are different depending on the animal's size. For instance, as animals get larger, they are increasingly likely to be injured by falls from great heights. When animals get much larger than, say, a rat, both their terminal velocity (maximum falling speed) and their momentum are relatively high, so the bigger they get, the more likely they are to be injured on impact with the ground. Gliding obviously reduces this risk and allows a falling animal to convert vertical speed into horizontal speed and achieve a safe landing.

Smaller animals, in contrast, have high aerodynamic drag due to their high surface-to-volume ratios, and also low body weight. These features in turn give them low terminal velocities and low momentum: a thumb-sized shrew or a pencil-sized millipede is unlikely to be injured by a fall from even a very tall tree. (A colleague who shall remain nameless dropped mice from the roof of a five-story building to demonstrate their low terminal velocity, and the mice seemed largely unaffected.) Scientists discovered the phenomenon of directed aerial descent (Chapter 3) in worker ants, who have almost no risk of being injured in a fall.[19] If you are arboreal and very small, the advantage of gliding is that you can avoid landing on the ground and can stay up in the tree. Most specialized tree dwellers are poorly adapted for getting around on the ground, and at least one study has shown that arboreal ants are much more likely to be eaten by predators on the ground than ants that normally live on the ground.[20] In short, gliders can better avoid landing on the ground where they are exposed to unfamiliar predators. This advantage does not just apply to small gliders; videos of captured colugos released on the ground show that they invariably hop as quickly as they can to the nearest tree, and they find it distressing to be on the ground at all.

Once a species evolves some gliding ability, another factor comes into play: energy conservation. A flying squirrel can glide from one tree to another distant tree using much less energy than a non-gliding squirrel would use climbing down the first tree, running across the intervening ground, and climbing back up the second tree. This difference is a great benefit when an arboreal animal needs to travel long distances or search many trees to find food. Not only does gliding require less energy but it is inherently faster. So a glider can search more territory faster and more economically than a similar non-glider.

## STEPPINGSTONE TO FLAPPING VERSUS THE SUCCESSFUL END STATE

Biologists are prone to refer to flapping flight as "true flight," thus relegating gliding to some other state of less-than-true flight. As I have said before, this distinction is aerodynamically absurd because a flying squirrel's wing membrane is doing the same thing as the wing of a jet airliner, and no one would argue that the airliner is not truly flying. Nevertheless, given that prevailing attitude, biologists and paleontologists in the past have often seemed to treat gliding as a sort of transitional state, a temporary steppingstone on the way to evolving powered flight. Indeed, bats probably evolved from arboreal gliders, and at least a few scientists think that may be true for birds and insects as well. Treating gliders as if they are in transition on the way to some more "advanced" state, however, misreads both the evolutionary process and the evidence; today most scientists would consider such an attitude to be obsolete.

I hope this chapter has shown that gliding is actually a sophisticated adaptation that involves changes in anatomy, sensing, behavior, and judgment. In addition to the obvious structural modifications to produce some sort of wing, the glider needs acute vision to accurately see distant landing spots and rapidly approaching obstacles. A gliding animal also needs the mental ability to judge whether a potential landing site is within safe gliding distance and must have the behavioral adaptations that allow effective aerial steering. Effective gliding is thus a complex adaptation, one that provides substantial benefits even without being actively powered.

Gliding has evolved dozens of times in the animal kingdom, and we have no evidence that these gliders were short-lived transitional forms. Although the fossil record is not very good for modern gliders, evidence from molecular phylogenetics suggests that *Draco* split off from other lizards at least 35 million years ago and has probably been gliding most of that time.[21] The great diversity of fossil and living gliders, as opposed to only four lineages of powered flyers, suggests that gliding is a smaller evolutionary step than flapping flight.

Yet even though gliding is a perfectly viable lifestyle on its own, at least some powered flyers evolved from arboreal gliders. In the debates surrounding the evolution of bird flight, some scientists have expressed the view that "gliders don't flap, flappers don't glide."[22] Leaving aside the fact that many flappers do glide routinely, the phrase illustrates the misapprehension that gliding and flapping are two qualitatively different, mutually exclusive processes. As we saw in Chapter 3, gliding and flapping are in fact two ends of a continuum, and nothing physical or aerodynamic prevents a glider from taking advantage of minimal attempts at flapping.

**REFERENCES**

1. E. H. Colbert (1967) *American Museum Novitates.*
2. H. Tennekes (1996) *The Simple Science of Flight: From Insects to Jumbo Jets.*
3. R. W. Thorington Jr. and L. R. Heaney (1981) *Journal of Mammalogy.*

4. S. P. Yanoviak, M. Kaspari, and R. Dudley (2009) *Biology Letters*.
5. E. Frey, H.-D. Sues, and W. Munk (1997) *Science*.
6. P. L. Robinson (1967) *Science & Culture*.
7. E. H. Colbert (1970) *Bulletin of the American Museum of Natural History*.
8. J. A. McGuire and R. Dudley (2011) *Integrative and Comparative Biology*.
9. P.-P. Li, K.-Q. Gao, L.-H. Hou, et al. (2007) *Proceedings of the National Academy of Sciences of the United States of America*.
10. N. C. Fraser, P. E. Olsen, A. C. Dooley Jr., et al. (2007) *Journal of Vertebrate Paleontology*.
11. G. J. Dyke, R. L. Nudds, and J. M. V. Rayner (2006) *Journal of Evolutionary Biology*.
12. J. Meng, Y. Hu, Y. Wang, et al. (2006) *Nature*.
13. A. P. Russell, L. D. Dijkstra, and G. L. Powell (2001) *Journal of Morphology*.
14. J. J. Socha and M. LaBarbera (2005) *Journal of Experimental Biology*.
15. B. J. Stafford (1999) [Dissertation].
16. J. Davenport (1994) *Reviews in Fish Biology and Fisheries*.
17. F. C. Wiest (1995) *Journal of Zoology*.
18. U. M. Norberg (1990) *Vertebrate Flight: Mechanics, Physiology, Morphology, Ecology and Evolution*.
19. S. P. Yanoviak, R. Dudley, and M. Kaspari (2005) *Nature*.
20. S. P. Yanoviak (2010) in *Ant Ecology*.
21. M. P. Heinicke, E. Greenbaum, T. R. Jackman, et al. (2012) *Biology Letters*.
22. K. L. Bishop (2008) *Quarterly Review of Biology*.

# CHAPTER 5

# Insects

*First to Fly*

Long before amphibious vertebrates first tentatively squirmed onto land, insects had already colonized land and evolved fully powered flight. Indeed, insects had already achieved flight more than 150 million years before the first, most primitive dinosaurs appeared, and at least 160 million years before pterosaurs followed them into the air as the second group of powered flyers. We have ample fossil evidence that flying insects were well established and diverse over 300 million years ago, but the details of how they got there are more than a bit murky.

## OLDEST INSECTS

Scientists have discovered an abundance of winged insect fossils from as far back as 318 million years ago, but prior to that, the insect fossil record is exceedingly sparse. Rocks from about 407 million years ago have produced fossils of a single species of springtail (order Collembola).[1] Springtails are actually fairly common today, but they are all tiny—the largest is about half a centimeter (⅕ inch) long and most are very much smaller—and they live in soil and leaf litter, so hardly anyone other than entomologists encounter them. They are called springtails because of a tail-like appendage that they use like a catapult for leaping, but no springtail has anything remotely like wings. In the past, biologists considered springtails to be extremely primitive insects, but nowadays most biologists consider them not to be true insects but the closest relatives of true insects. In any case, those 407-million-year-old fossil springtails represent the oldest known fossils on the lineage leading to insects. Other rocks from about 380 million years ago yielded fragments of a bristletail (order Archaeognatha),[2] a very primitive type of insect that looks superficially like a silverfish.* Both the springtail and the bristletail diverged from the lineage leading to winged insects well before flight evolved; scientists call such animals "primitively wingless" as opposed to insects descended from flying insects that have secondarily lost wings. Until recently, scientists had nothing to link these most ancient insect and related fossils with 100-million-year-younger flying insects.

Sometime around 2000, entomologist David Grimaldi and my University of Kansas colleague, Michael Engel, were on a trip to visit museums around the world. They were writing a book on the evolution of insects, and they were visiting museums to take original photos of fossil insect specimens to illustrate their book. As Michael recalls, they were on the way back from Moscow and decided to make a brief stop in London to look at the original fossils of those 407-million-year-old springtails at the British Museum of Natural History. While they were looking at the microscope slides of the collembolans, Michael happened to notice another slide in the same drawer of a different specimen from the same fossil formation. (It had been given a name, *Rhyniognatha*, more or less in passing when it was first prepared in the 1920s, but it was never really described.) Deciding he might as well take a quick look at it since they were there anyway, Michael put it on a microscope—and was stunned. The fossil appeared to be a set of mandibles or jaws with a double hinge ("dicondylic") that is utterly diagnostic of flying insects. In other words, if it had been from a living insect, it would have unquestionably come from an insect with wings. Trying not to show too much excitement, Grimaldi and Engel casually asked if they could borrow this "nondescript" specimen to take it back to the University of Kansas for more careful study. When they published their study, scientists realized that this fossil could mean that insect flight had already evolved by 407 million years ago.[3] Unfortunately, all that survived were the mandibles; no trace of the thorax fossilized, so we don't know for sure that this insect had wings. Moreover, as Michael has explained many times, the conditions (boiling mineral springs) where the fossils formed would most likely have destroyed wings if they had been present, so he says looking for wings in the flint-like rocks where the mandibles were found is "a fool's errand." We thus have a tantalizing hint that insect flight may have evolved nearly

---

* Despite the name, silverfish are insects, not fish. They have a silvery appearance due to a covering of very fine scales, and their bodies do have a vaguely fishlike shape.

100 million years earlier than previously assumed, which would make the gap between the time of the origin of flight and the oldest known fossils of winged insects even longer.

Scientists face a major puzzle in trying to understand the evolution of wings and flight in insects. Starting about 318 million years ago, the fossil record becomes rich in winged insects (including many isolated wings). Because insect wings usually have distinctive vein patterns that identify their lineage, these early insect wing fossils clearly demonstrate that a lot of evolution and diversification had already occurred among flying insects. For all that diversification to have already happened, flight must have first evolved very much earlier—many tens of millions of years at a minimum. This observation, combined with Engel's new look at those tantalizing mandibles, led Engel and his colleagues to suggest that insects must have evolved flight at least 400 million years before the present (and possibly considerably earlier), but so far we have found no insect fossils of any kind earlier than those frustratingly enigmatic jaws from 407 million years ago, nor any trace of insect wings older than 320 million years ago. This gap in the fossil record might seem implausible, but due to tectonic changes of the continents, terrestrial and freshwater fossils from that period are extremely hard to find. Early insects were most likely terrestrial,[4] and since the main source of fossils of terrestrial animals is from those that fall into swamps and get buried in silt, the absence of non-marine fossils from that time means that we have no useful source of insect fossils from the relevant periods (late Silurian through mid-Carboniferous, 420 to 320 million years ago). We thus have a weak suggestion of when insects may have evolved flight, but so far, we have no direct evidence of any of the early stages of wing evolution in insects.

Just to add a bit to the puzzle, in 2014 a German-Czech team led by Arnold Staniczek re-described a 309-million-year-old fossil, originally described as a juvenile mayfly, as a wingless adult. *Carbotriplurida kukalovae* was a large insect (10 centimeters or 4 inches long) with long legs, big eyes, and long tail filaments. It also had flat, side extensions of the roof of the thorax and abdomen just as predicted by the paranotal lobe theory described later in this chapter. These researchers suggest that *Carbotriplurida* was evolutionarily intermediate between the wingless Zygentoma (silverfish and relatives) and the winged insects.[5] *Carbotriplurida*, however, lived many tens of millions of years *after* insects had already evolved flight, so if it does represent the direct ancestor of winged insects, a wingless part of the lineage must have split off early and persisted remarkably long into the age of winged insects.

## TRADITIONAL WING ORIGIN THEORIES

A lack of informative fossils has not prevented scientists from speculating about the evolutionary beginnings of insect flight. The arguments about possible wing precursors go back well over a century. In fact, the two traditional, widely accepted theories were both first described in an evolutionary context in the mid-1800s. In 1870, Carl Gegenbaur converted a 60-year-old opinion that insect wings were modified from larval gills into a more modern scientific theory consistent with Darwin's just-published theory of evolution.[6,7] Almost immediately, in 1873, Fritz Müller proposed the alternative paranotal lobe theory (see below) as a counterargument because he thought some developmental evidence in termites was inconsistent with the gill theory.[8,9] These two theories were the only widely accepted theories to explain the evolution of insect flight throughout the 20th century. Even after all that time, variations and updates of these theories are still accepted by many scientists to this day.

### The Gill Theory

The gill theory (also known as the pleural appendage theory) is based on the fact that several lineages of insects have juvenile stages that possess several pairs of external, movable gills, usually on the abdomen (Fig. 5.1). Not only are these gills movable but they often contain tracheae (air tubes) in a branching pattern reminiscent of wing veins. Proponents of the gill theory suggest that the ancestors of winged insects had juveniles with gills on the thorax as well as the abdomen. Moreover, some modern aquatic insect larvae (technically naiads; see Box 5.1 for terminology used for immature insects) use these gill plates more like fans to drive water over additional respiratory surfaces, and some even use them as paddles for swimming. The main idea of the gill theory is that some ancestral insect evolved to use the gill plates on the thorax more as paddles than for respiration. Then if this insect spent part of its time out of water, the thoracic paddles might have allowed steering during leaps or falls, or a little gliding to extend leaps, or even weak flapping to aid steering and gliding. As the thoracic paddles became useful in air, natural selection would have led to bigger paddles to produce more lift and thrust and more effective steering, and eventually to stronger muscles and more refined flapping movements to achieve fully powered flight.

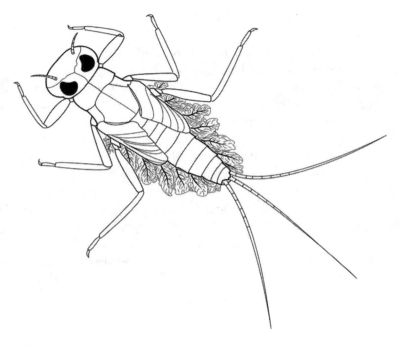

**Figure 5.1:**
An immature mayfly ("naiad"), showing the rows of gills along its abdomen. (Courtesy of S. T.)

---

*Box 5.1:* IMMATURE INSECTS

The terminology for juvenile or immature insects can be challenging. Entomologists call an immature individual of a species that goes through a pupal stage (those with "complete" metamorphosis) a "larva," plural "larvae." Caterpillars, fly maggots, and beetle grubs are familiar examples. Outside of entomology, a "larva" can be any immature stage, although it usually refers to a juvenile that looks different from the adult; non-entomologist biologists thus tend to call all immature insects "larvae." Juveniles of insects without a pupal stage ("incomplete" metamorphosis) tend to look rather like miniature adults although without wings or external reproductive structures, so they are properly called "nymphs." Aquatic juveniles of insects with incomplete metamorphosis—for example, dragonflies, mayflies, or stoneflies—can be considered nymphs, but since they look noticeably different from the adult stage, entomologists often refer to them as "naiads."

> **Box 5.1: Continued**
>
> Curiously, "nymph" and "naiad" are both borrowed from Greek mythology, Nymphs being minor female deities associated with specific places and Naiads being a subset of Nymphs associated with streams and lakes. Using "naiad" has a certain logic to it, but using "nymph" makes no sense to me, especially given that Nymphs are often associated with liberated sexuality. Insect nymphs are, by definition, incapable of reproductive behavior!

The gill theory has the advantage of starting with a movable appendage in more or less the right place on the body, a useful exaptation for steering and incipient flapping. This theory has some drawbacks, however. First, if these gill-paddles served a partly respiratory function underwater, they would be unable to do so effectively in air. In order to have a thin enough covering to allow sufficient gas exchange under water, gills are usually too floppy to support themselves out of water, typically collapsing into ineffective, low-surface-area clumps in air. Most gills also require some ventilating mechanism to drive a flow of water over them, and a pump evolved to move water would be useless in air. Second, simply passing through the air-water interface can be quite difficult for an insect-sized animal. Many insects depend on the surface tension of the air-water interface to support their entire body weight, water striders being a familiar example. The fact that an insect's entire body weight is not enough to push through the surface shows how much effort tiny animals need just to pass through that interface. The large surface area of thoracic gill-paddles big enough to be useful in air would make exiting water even more difficult if not impossible. The biggest problem, however, is that most entomologists think that early insects evolved on land, and those with aquatic larvae evolved secondarily from terrestrial ancestors.[4] The reason is that the insect respiratory system consists of a branching network of air tubes throughout the body called tracheae. Tracheae can be quite effective at exchanging oxygen and carbon dioxide in air, but this system simply does not work well underwater and certainly would not have evolved there. The tracheal system is a fundamental feature of insects and it unquestionably evolved in air. Bristletails and silverfish are members of two ancient lineages that split off from other insects before wings evolved and they both have tracheae, so most entomologists think that the earliest insects must have been terrestrial.[10]

## The Paranotal Lobe Theory

The essence of the paranotal lobe theory is that the ancestors of winged insects evolved flat plates extending from the top of the thorax out to each side. These are the paranotal lobes, "para-" meaning "alongside" and "notum" being one of the plates of the exoskeleton that make up the roof of the thorax. The main idea here is that this ancestral insect could have evolved these lateral extensions for some non-aerodynamic function—for example, camouflage, side protection, or strengthened anchoring for leg muscles—although an early aerodynamic function cannot be ruled out. Whatever the original use, these plates or "lobes" eventually became large enough to have a significant aerodynamic effect. They might then have been used for parachuting, or for gliding to extend leaps from foliage or other elevated perches. These lobes would initially have been fixed or immobile. If, however, they could be adjusted for steering, they would have been more versatile for gliding or just steering during leaps. They would then have evolved a joint or hinge, which in turn would require co-opting some muscles to control their movements. Once the lobes became movable for steering, the basic building blocks were in place. The lobes would get larger to produce more lift, and the steering movements would evolve from weak flapping to fully powered flight.[11]

The main objection to the paranotal lobe theory is that it requires the incipient wing to evolve a new joint or articulation from scratch. Although not impossible, evolution of completely new joints in rigid skeletons is surprisingly rare; fusion of skeletal elements (bones, or plates of exoskeletons) or loss of joints seems fairly common, but developing an entirely new joint within a single plate or bone, as opposed to modifying an existing joint, is extremely unusual. Most books and articles that discuss the paranotal theory, however, mention one thought-provoking example of a new joint: that in oribatid mites. Mites are relatives of spiders, not closely related to insects, and the oribatid mite lineage has evolved what amount to paranotal lobes, including a new hinge, although these paranotal lobes have no known flight-related function.[10]

For most of the 20th century, the paranotal lobe theory was more widely accepted than the gill theory. For example, Richard Alexander reviewed previous work and argued that the paranotal lobe theory was more likely. Noting that only adults have wings, he suggested that paranotal lobes may have been used originally in courtship displays or mating behavior.[12] If lateral thoracic extensions had originally evolved as courtship displays to attract members of the opposite sex, sexual selection* can indeed be a potent agent to increase elaboration, as amply demonstrated by peacock tails. Sexual selection could have driven selection for ever-larger lobes and even conceivably a new joint allowing movement to increase the effectiveness of the display. Alexander contended that if an insect with such advertising devices were also an active, vegetation-inhabiting species, its thoracic display structures could also help guide and extend leaps within and between plants so that the insect could avoid landing on the ground.

Although the paranotal lobe theory probably spent more time as the leading contender, proponents of the gill theory never disappeared. In 1973, eminent insect physiologist Vincent Wigglesworth revised and updated the gill theory by suggesting that wings evolved from side branches off the bases of legs called "exites" or "exopodites." He pointed out that very primitive insects like silverfish have tiny, peg-like appendages called "styli" on the thorax and abdomen, which he suggested were vestigial remnants of larger, mobile exites. (See Box 5.2. Insect and Crustacean Limbs.) If exites had evolved into gills in aquatic species, they would have provided the basis for mobile protowings,[13] although they would have had to shift up higher on the side of the body somewhere along the line.

Jarmila Kukalova-Peck, a prolific insect paleontologist, took Wigglesworth's suggestion a step further. She suggested, based on fossils, that the primitive insects had miniature legs on all the abdominal segments as well as large legs on the three thoracic segments. She argued that the primitive insect leg had an extra basal segment that evolved into the side walls of the thorax and abdomen, and the exites of this hypothetical segment formed flattened, mobile plates all along the upper sides of the thorax and abdomen. These exites all evolved into gills in an early aquatic lineage, and from there, her suggestion follows the standard gill theory. Although she assumed that an aquatic stage was part of the process, her exites could in fact fit equally well into the terrestrial scenario as either fixed or mobile paranotal lobes. Kukalova-Peck seems to see more detail in fossils than other researchers, and few other scientists accept her hypothetical leg segment; so until better fossils come along, her version of the gill theory remains controversial.[5,14,15]

---

* Sexual selection is a form of natural selection (first described by Charles Darwin) where one sex chooses mates based on some attribute not otherwise related to survival, which then causes that attribute to become exaggerated. The bright red color of the male cardinal is another example.

> *Box 5.2:* INSECT AND CRUSTACEAN LIMBS
>
> Crustacean limbs are surprisingly complex compared to insect limbs. Crustacean limbs are based on a forked or Y-shaped main axis of several segments, often with smaller side branches (exites and endites) as well. Biologists call such limbs "biramous." For instance, on the lobster thorax, the inner or medial main branch (endopod) develops into a walking leg and the outer or lateral main branch (exopod) becomes a feathery gill hidden under the carapace. An exite is a small, lateral side-branch from near the base of the limb.
>
> Insect legs, in contrast, are unbranched, forming a single column made of several segments. Insects share these so-called uniramous limbs with myriopods (centipedes and millipedes); for a long time, biologists thought that this trait indicated a close relationship between insects and myriopods. "Uniramia" was even used as a name for the group uniting the two. In the last few years, however, the evidence from molecular phylogenies has become quite convincing that crustaceans are the closest relatives of insects. In fact, insects are nested within the crustacean phylogeny and so are technically a subgroup of crustaceans. This means that insects and myriopods evolved their uniramous legs independently. Wigglesworth's suggestion was controversial at the time because insect legs were then considered to be fundamentally unbranched. Given the evolutionary relationship between crustaceans and insects, Wigglesworth's contention that styli of silverfish may be vestigial exites cannot be discounted. Whether that has anything to do with the evolution of wings is a separate question.

## MODERN THEORIES: EXPERIMENTS AND OBSERVATIONS
### Model Tests

Frustrated by the lack of fossil evidence for wing origins in insects, a number of researchers have taken a different tack. Some have used experiments to try to tease apart the physical features and constraints that might have favored acquisition of wings. Others have looked at living insects to find behaviors that might have a bearing on the evolution of flight.

For example, evolutionary biologist Joel Kingsolver and biomechanics specialist Mimi Koehl built a series of models of hypothetical insects to test their thermodynamic and aerodynamic properties. Earlier scientists had looked at insect wings as possible solar collectors—warm muscles contract faster, which might be favored by natural selection—or at aerodynamics of cylinders as simplified stand-ins for wingless insect bodies.[16,17] But Kingsolver and Koehl took a more comprehensive approach. They wanted to see if small, flat thoracic projections or "protowings" could have provided any non-aerodynamic benefits and to figure out the conditions where such protowings might provide aerodynamic advantages. They used models over a range of sizes from 2 to 10 centimeters (about 1 to 4 inches) in length and either wingless or with wings over a range of lengths (Fig. 5.2A). They looked at thermal properties by measuring the models' core temperature under flood lamps and their aerodynamic properties in a wind tunnel.[18] Their results show intriguing interrelationships.

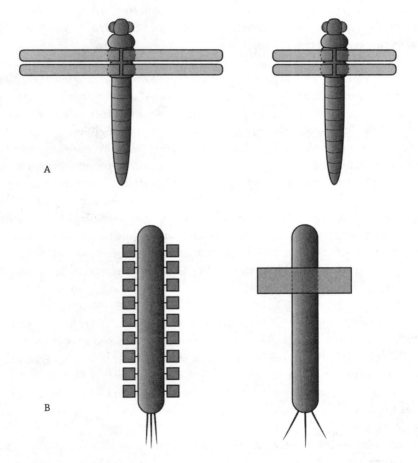

**Figure 5.2:**
Simplified models used to test theories of insect wing origins (not to scale). A. Models used to compare thermal and aerodynamic properties, redrawn from Kingsolver and Koehl (by permission of Cambridge University Press).[18] B. Models used to test gliding stability, redrawn from Wootton and Ellington (by permission of John Wiley & Sons).[20]

They found that stubby wings are aerodynamically effective—increasing glide distance—on big insects, but small insects need rather long wings to get similar benefits. Conversely, short, stubby winglets make very effective solar collectors, but longer wings provide little or no additional solar warming. The effect was most pronounced in the largest insect model; here stubby winglets of less than a ½ centimeter (⅕ inch) greatly improved solar heating over no winglets, whereas the smallest model gained little or no extra heating from short winglets. Kingsolver and Koehl concluded that if natural selection favored solar heating, perhaps to warm up faster than your predators or prey on cool mornings, stubby thoracic projections on medium- to large-sized insects could be advantageous. If such projections could be adjusted to orient toward or away from the sun, they might have been even more effective, and this would have favored movable plates.

Kingsolver and Koehl showed that if such a hypothetical ancestral insect were large, its solar-collectors-cum-protowings would not have had to get very big to start having some aerodynamic effects. Small insects, in contrast, would need quite long protowings to get much aerodynamic benefit. With big enough protowings, all the advantages of gliding (Chapter 4) would come into play, particularly if the solar collectors were movable to allow steering. If this insect benefited from increasing glide distances, then increasing wing size and developing flight muscles would have been favored—and this would have put such an insect well on the road to flapping flight. Some scientists have called this analysis oversimplified, saying that such solar collectors would be quite cumbersome when not in use and that not all situations would necessarily favor heated muscles. Nevertheless, this study showed how thermal biology and aerodynamics could interact to favor wing evolution.

In a later review article, Kingsolver and Koehl referred to this approach as "bounded ignorance,"[19] which is one of my favorite technical terms. What they meant is that we can use experiments to show what is physically possible (or impossible) and what factors could be advantageous—placing boundaries—even when the gaps in the fossil record mean that we remain ignorant of what actually happened. A number of other studies have taken this bounded-ignorance approach to the study of flight evolution.

British zoologist Robin Wootton and his colleague Charles Ellington (a top expert in insect aerodynamics) performed one of these studies. They also built models, which they tossed as gliders to study stabilizing effects of a variety of possible anatomical embellishments. Much earlier, aeronautical engineer J. W. Flower had shown that insect-sized cylinders could achieve surprisingly good glide angles (up to 45 degrees) if they could be stabilized at the proper falling angle.[17] Wootton and Ellington tested model cylinders with either a series of nine pairs of small adjustable winglets down each side or a single pair of larger, fixed winglets near the front, each version with and without long tail filaments (Fig. 5.2B). Their main goal was to see whether these structures could give the cylinders the necessary stability to fall at an angle that generated lift and, hence, horizontal motion.[20] In other words, could they achieve a decent glide angle with a basically cylindrical body? Their answer was a qualified "yes," depending on the particular combination and orientation of the structures. For example, the glide angle and stability of the model with several pairs of small winglets was largely set by the angle of the hindmost pair of winglets, and the model became unstable if that pair was removed. By giving the model paper "tail" filaments like those of many primitive insects, however, the model regained stability. These scientists thus showed that a hypothetical ancient insect could have converted a vertical fall into a 30-degree to 45-degree glide with small, stubby winglets and tail filaments. As with other researchers, they found that the aerodynamic benefits were greatest for models representing larger insects.

Wootton and Ellington thus showed that if the ancestor of winged insects had a series of little plates down the side of the thorax and abdomen (as in the gill theory), and if the plates were adjustable, the plates did not have to be very big to provide some gliding. Alternatively, they showed that a single, immobile, stubby pair of winglets (as in the paranotal lobe theory) along with a set of tail filaments was not quite as effective but still gave the cylinder the stability needed to convert a fall to a glide. So even very short, fixed wing surfaces could have provided some aerodynamic benefit.

All of these model studies can help explain what is physically possible and what is improbable, but they cannot produce a definite answer to the question of how wings evolved. Moreover, their results are quite size-dependent. The effectiveness of wings can change quite a bit over the range of size from small insects to the largest of these models (16 centimeters, about 6 inches) because wing performance is quite sensitive to a scale factor called the Reynolds number (see Box 3.1). If the direct ancestors of flying insects were smaller than 1 or 2 centimeters (½ to 1 inch) in length, most of these model studies would not have accurately described the aerodynamics. If the pre-flight insects were moderately large (in the 6- to 10-centimeter or 2- to 4-inch range or bigger), as some scientists suggest, then the model studies could be relevant. So far, however, we have little or no direct evidence of the size of these early insects.

## Modern Analogs

A very different method from building physical models is to look at structures, processes, and behaviors in living animals and imagine how similar features might have assisted the evolution of flight. This is still a form of bounded ignorance: practitioners don't say that living animals inherited the relevant features directly from the earliest flyers; rather, they propose that if modern animals could evolve such features, so might have the ancient insects who first evolved flight. James Marden's surface-skimming theory, stimulated by his observations of living stoneflies, illustrates this approach.

Stoneflies are rather run-of-the-mill, unspecialized insects with aquatic nymphs (larvae) and air-breathing adults. Even though most adult stoneflies can fly, they are not particularly strong flyers so they rarely stray far from water. Evolutionary biologist James Marden noticed that on cold days, some stoneflies could not summon the power to fly up in the air, but they could use their wings as propellers to skim quickly over the surface of the water like an Everglades airboat. He and his then-student Melissa Kramer studied the mechanics of surface skimming. They found that they could clip off up to 80% of a stonefly's wing area, and the stonefly could still locomote across the water surface. Skimming speed, however, decreased as more and more of the wing was removed. Marden and Kramer concluded that surface skimming might be an intermediate stage in the evolution of flight.[21] If an aquatic insect with gill plates (or gill-like paddles) evolved an air-breathing adult stage, it might be able to skim across the water surface even with small, weak winglets. Selection for faster skimming would lead to longer wings and faster muscles. These improvements might then allow hops off the water at first, leading eventually to powered flight. In later work, Marden pointed out that some stoneflies don't even flap their wings, they just raise them up and angle them to catch the wind, exactly like a sailboat.[22] That behavior pushes the theory even further back, to a semi-aquatic insect that used gills or paddles as sails when on the water surface.

Marden has suggested that surface skimming may be a relict behavior directly inherited from the ancestral insects that first evolved flight.[23] Most scientists, however, think skimming stonefly behavior evolved long after the origin of flight rather than having been directly inherited from distant ancestors. Yes, stoneflies are fairly primitive, but dragonflies and mayflies are even more primitive, so the ancestors of stoneflies must surely have been capable of effective, fully powered flight; surface skimming evolved secondarily from powered flight in stoneflies (as well as in some flies and caddisflies). Moreover, as I have said, the preponderance of evidence is that ancient insects were terrestrial, not aquatic.[4,10,24]

In my view, the most significant description of a feature of modern animals relevant to the evolution of insect flight, perhaps even animal flight in general, is work by Steven Yanoviak, Robert Dudley, and Michael Kaspari on directed aerial descent in ants (as discussed in Chapter 4).[25-27] The possibility that many arboreal animals can direct their descent without any overt aerodynamic modifications means that we need to reconsider many of the arguments about the origins of gliding and flying animals. Yanoviak and colleagues' work was actually foreshadowed in an article by German entomologist I. Hasenfuss. Hasenfuss started from the premise that the closest non-flying relatives of the winged insects are those in the order Zygentoma (silverfish are common members of this group). He assumed that the ancestors of flying insects must have had anatomy and lifestyles similar to silverfish and their kin (Fig. 5.3): fast runners with flattened bodies to fit easily into cracks and crevices. Hasenfuss dropped silverfish from various heights and observed that they have a cat-like ability to always land on their feet, even if dropped from just a few body-lengths above the surface. They spread their legs, antennae, and tail filaments to slow their fall and bend their abdomens to roll over so that they always land right-side-up, ready to scuttle off.

Hasenfuss then made simplified models of silverfish and dropped them to see if they could glide.[28] He found that the models fell with the head aimed slightly down in a stable, steep glide. He then added short, thoracic winglets to his models. He found that even with very short winglets, he could more or less double glide distances of larger models; his small models, however, did not glide noticeably better with short winglets, paralleling what Kingsolver and Koehl had found.

Hasenfuss hypothesized that the ancestral, silverfish-like insect ate at the tips of growing plant shoots. If so, this insect could save time and energy by jumping off the plant after it finished eating the most tender

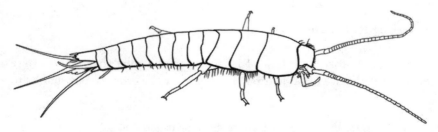

**Figure 5.3:**
A silverfish, a primitively wingless insect. (Courtesy of S. T.)

part, landing safely, and running to climb up the nearest uneaten plant. Some modern spiders can also reorient in falls to land on their feet, and spiders were probably common predators of these ancient insects. If so, then the insects would have benefited if they could angle their fall a bit so they could land a little farther from the original plant (and from a pursuing spider) and closer to a new plant. Indeed, parachuting to slow the fall could even be detrimental, so a fast, steep glide would be most useful in this setting. Natural selection would favor any plate-like extensions of the thorax to extend glides and eventually reach distant plants without spending any time on the ground at all. Hasenfuss even suggested how flappable wings might have evolved based on quite sophisticated anatomical comparisons and described how initial weak flapping would have aided escape from predators.[28]

If, as Yanoviak and company suggest, many animals prone to jumping or falling from substantial heights can direct their descent toward the nearest tree, then the behavioral exaptation to steer a glide might well be common in arboreal animals. Ants are neither primitive nor ancestral to other flying insects, but they are wingless* and nothing about their anatomy hints at any aerodynamic function. The ability of some ants to predictably steer a fall demonstrates that any arboreal animal, whether it has obvious gliding surfaces or not, may have the sensory and behavioral ability to steer a glide. Moreover, additional studies have shown that a wide variety of flightless animals use some aerodynamic behaviors (righting, steering).[29] If natural selection strongly favored longer glides, such an animal would tend to evolve enlarged surfaces to improve lift production. This selective force could come from predator evasion, quicker movement to new trees, avoidance of time on the ground, or some combination. "Parachuting" for soft landings or wind dispersal would probably not have been one of these selective forces (contrary to arguments of earlier researchers)[30,31] because parachuting favors very small body size and high drag, which work against the evolution of both gliding and flapping.[10,18]

### A Working Hypothesis

Due to the unfortunate lack of fossils from the time that insects were evolving flight, we have no direct evidence for the process; but thanks to

---

* Worker ants, who make up the overwhelming majority of colony members, are entirely wingless. Reproductive individuals (drones and queens) have wings before mating; males die and females shed their wings after mating.

the bounded-ignorance approach and based on processes in living insects, we can make some fairly sophisticated guesses about how insect flight probably evolved. The direct ancestors of flying insects were probably rather silverfish-like and probably at least a centimeter (half inch) or so long, perhaps much bigger. They probably fed on the reproductive structures or growing tips at the tops of stalk-like plants and routinely jumped to the ground to escape predators or to speed up travel to nearby plants. They probably had fairly good vision (unlike modern silverfish, but like some fossil relatives) to see approaching predators and find new plants to feed on.

This ancient species surely had the ability to reorient and land right-side-up during a jump or fall and may well have already evolved the ability to steer its fall back toward its original plant or toward a new one. It may have had lateral extensions—either fixed or movable—along its thorax and abdomen for some non-flight function, perhaps like those of *Carbotriplurida*. If these extensions produced even a little bit of lift, they could extend the jump into a glide; and under the right conditions, they could evolve larger surface area and eventually become wings. If the extensions were already hinged, flapping may have evolved almost simultaneously; if not, leg muscles might initially have deformed the exoskeleton of the thorax enough to move the wings a bit, even without a flexible hinge. A hinge would have evolved eventually to allow greater wing movement for steering and flapping. Flapping movements would have increased in strength and complexity until the insect achieved fully powered flight.

## FOSSIL EVIDENCE: EARLIEST KNOWN FLYERS

As we have seen, scientists discovered a few fossils from very primitive insects that lived over 400 million years ago, then nothing for the next 100 million years.* Rocks from just over 300 million years ago have yielded many fossils of insects and especially insect wings. Frustratingly, these fossils are mostly from insects clearly capable of fully powered flight. Thus, while primitive in many ways, they were well beyond the transition from winglessness to powered flight.

---

\* Researchers recently described a fossil from an ancient swamp in what is now Belgium that appears to be a 360-million-year-old juvenile (hence, wingless) insect from the winged insect lineage.[32] This fossil could help pin down the age of the lineage but is no help in clarifying wing evolution. Other scientists, however, see more crustacean than insect characteristics,[33] and the lone specimen is not well enough preserved to convincingly resolve this dispute.

These ancient insects include a number of extinct groups that were once widespread and common, and they also offer some clues about a major wing innovation that separates most modern insects from their more primitive ancestors and living sister groups (see phylogeny, Fig. 5.4). This innovation is the neopterous wing articulation, a modification that allows most living insects to fold their wings flat over their abdomens at rest by flexing them backward, as in modern house flies and bees. More primitive flying insects have a simpler wing hinge, called a paleopterous articulation, that allows the wings to flap up and down but not to fold flat over the abdomen. Dragonflies, for example, have paleopterous hinges, so they perch with the wings outstretched when at rest. Scientists used to consider insects with the paleopterous hinge to be a single lineage (called Paleoptera), but now generally consider them to represent several different lineages. Insects with the neopteran wing hinge do form a monophyletic group, according to both anatomical and genetic evidence, so they are placed together in a taxonomic group called Neoptera.[4]

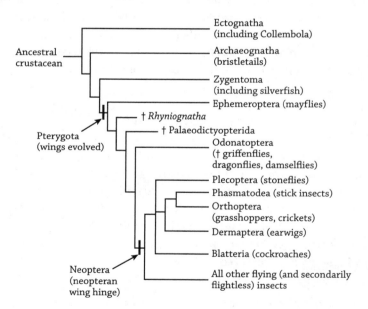

**Figure 5.4:**
Simplified phylogeny based on Grimaldi and Engel.[4] (Several lineages omitted for clarity.) A vertical bar represents the base of the indicated lineage (or first appearance of a trait characteristic of that lineage); all subsequent branches beyond—to the right—of the bar are part of that lineage. †: Extinct.

## Palaeodictyopterida

The Palaeodictyopterida was a major lineage of insects with paleopterous wing articulations that flourished early and then went extinct at the end of the Permian (about 250 million years ago).[34] From the time they first appeared, well over 300 million years ago, until the great End-Permian Extinction, Palaeodictyoptera made up about half of all insect fossils.[4] These insects had prominent beaks that they apparently used to pierce and feed on plant tissues. Some fossils even preserve distinct banding and other color patterns on the wings. Many of them had a curious little third pair of wings ("prothoracic lobes" or "winglets"), complete with veins, in front of the two normal pairs of wings (Fig. 5.5). Scientists are divided on whether these prothoracic winglets were

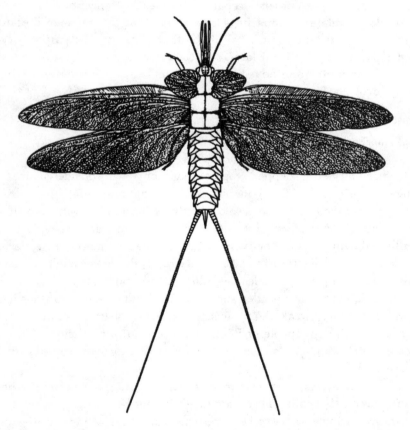

**Figure 5.5:**
Reconstruction of paleodictyopteridan, from Kukalova-Peck, used by permission of John Wiley & Sons.[32]

actually mobile, for steering or flapping, or fixed, perhaps as some sort of stabilizer or protective cover.[4] Scientists have identified juveniles of palaeodictyopteridans and these all appear to be terrestrial.[35-37] Most researchers consider the fact that these nymphs were air-breathers to be convincing evidence that the ancestors of winged insects were terrestrial and not aquatic.

Two other lineages of insects share the paleopterous wing articulation with palaeodictyopteridans: Odonatoptera, including modern damselflies and dragonflies and their extinct relatives, and Ephemeroptera, consisting of fossil and modern mayflies. Because these three lineages possess a primitive wing hinge and are present among the earliest winged insect fossils, they were traditionally considered to be three branches of a single primitive lineage called Palaeoptera.[38,39] Recently, biologists and paleontologists have uncovered mounting evidence that these groups represent three separate lineages that may not be particularly closely related. A consensus is emerging that the ephemeropterans are the most primitive or basal of the flying insects, the palaeodictyopteridan lineage split off later, and the odonatopterans were the last group to split off from the rest of the insects before the neopteran wing hinge evolved (Fig. 5.4).

### Ephemeroptera and the Subimago

Mayflies, in the order Ephemeroptera, have long been considered among the most primitive of flying insects. Their paleopterous wing hinge obviously places them near the base of the flying insect phylogeny, but they have another, uniquely primitive trait: they molt once after they develop full-sized wings. In all other winged insects, the last molt an insect ever makes is from a flightless immature stage to a fully winged adult. In mayflies, the last aquatic juvenile stage fills with air and floats to the surface where it molts to the so-called subimago (subadult) stage. This stage has full-sized wings, weak flight muscles, and undeveloped reproductive structures. Barely capable of flight, it flutters quickly to shore, finds a secure perch, and molts almost immediately to a fully flight-capable, fully mature adult.

Many scientists consider the mayfly subimago stage to be primitive because insects that split off the main insect lineage before wings evolved—primitively wingless insects like silverfish—continue molting and growing after reaching sexual maturity. In contrast, no flying insects other than mayflies molt after developing functional wings. There is a very good reason for this difference: in all other winged insects, the living tissue needed to form a new wing exoskeleton dies and is reabsorbed in the wing membrane, remaining only in the wing veins (the epidermis, see Fig. 2.4). Without living cells in the wing membrane, the insect can't form a new membrane exoskeleton and thus can't molt. In contrast, the mayfly subimago retains living tissue in its wings; this allows molting but makes subimago wings heavier and less rigid than a normal adult wing. Few scientists were surprised when molecular genetic analyses placed mayflies at the very base of the flying insect phylogeny. These insects apparently separated from the rest of the winged insects before flying insects lost the ability to molt as adults.

Mayflies have a rich fossil history, with many specimens of juveniles and adults described, including several extinct families from the Permian (over 260 million years ago). Modern mayflies have small, sometimes nearly vestigial, hindwings, and the adults have incomplete, nonfunctional digestive systems. The adult stage rarely lasts more than 24 hours, so adults do not eat. Fossils show that some of the early mayflies had hindwings as big as forewings and apparently functional jaws. A few of these ancient mayflies were huge (see Box 5.3. Carboniferous Giants);[40] most, however, were of modest size. By the end of the Jurassic (approximately 150 million years ago), mayflies had evolved to their modern form, and some fossil species can even be placed into families of living mayflies.

## Odonatoptera and the Griffenflies

The remaining insect group with the paleopterous wing hinge is the Odonatoptera, the group containing dragonflies and damselflies as well as the fascinating, extinct griffenflies (formerly called "giant dragonflies," a misleading name because griffenflies were a separate lineage from true dragonflies). This lineage includes the largest known insect of all time, the griffenfly *Meganeuropsis permiana* (see Box 5.3. Carboniferous Giants), with a wingspan of nearly 70 centimeters (28 inches)! Griffenflies, particularly the big ones, are known mostly from fossils of isolated wings or wing fragments (for example, see Fig. 5.6).[41] Based on a few partial body fossils of smaller species, griffenflies seem to have been aerial predators, much like modern dragonflies, but their wings lack some diagnostic features of true dragonfly wings. Dragonfly and damselfly fossils first appear in the early Permian, some 10 or 20 million years after the griffenflies.

> *Box 5.3:* CARBONIFEROUS GIANTS
>
> In the late Carboniferous, roughly 300 million years ago, a variety of terrestrial animal groups evolved enormous body sizes. For example, some amphibians reach a length of over 6 meters (20 feet). The largest terrestrial arthropods of all time lived then, including *Arthropleura*, a centipede that reached 2.6 meters (over 8 feet) in length, as well as 1-meter (3-foot) millipedes and 70-centimeter (28-inch) scorpions. The largest known flying insects also lived during this period.
>
> Scientists think gigantism evolved because the proportion of oxygen in the atmosphere was higher during much of that time.[38] Some geophysical studies suggest that atmospheric oxygen rose as high as 35% (compared with our current level of 21%) of the total atmosphere.
>
> More atmospheric oxygen would have let animals distribute oxygen to tissues over greater distances in their bodies, allowing larger body size. This improvement would have been especially significant for insects, with their diffusion-powered tracheal respiratory system.
>
> Although most carboniferous insects were no larger than modern-day insects, many were quite large and a few were immense. Curiously, many of these giant flyers are known only from fossils of isolated wings or wing fragments (for example, see Fig. 5.6), and many of these are from the Elmo fossil bed in central Kansas, a couple of hours drive from my office at the University of Kansas. Because researchers from Yale and Harvard did most of the original collecting at Elmo, however, the vast majority of the fossil insect specimens from Elmo are now housed in museums at those universities.
>
> These Carboniferous giants dwarf any flying insects alive today. Some mayflies had wingspans of over 43 centimeters (17 inches) and several species of palaeodictyopteridans had wingspans of over 50 centimeters (20 inches). Griffenflies, members of the Odonatoptera, include the largest known insect of all time, the griffenfly *Meganeuropsis permiana* with a wingspan of 66 to 70 centimeters (26 to 28 inches). Some of the "smaller" Kansas griffenflies include *Megatypus schucherti* with a wingspan of a mere 40 centimeters (16 inches) or *Typus gracilis* at 33 centimeters (13 inches), still impressive by modern standards.[39]

## WING NUMBER

The four-winged arrangement seems to have been the original pattern from which all other insect wing arrangements are derived. (We will ignore the prothoracic winglets of some fossil species because they are

**Figure 5.6:**
Griffenfly wing traced from a fossil (courtesy of Roy J. Beckemeyer, used with permission). Inset: Tracing of a wing of *Anax junius*, one of the largest living dragonflies, for scale (courtesy of S. T.).

quite small relative to the main wings, and we don't even know for sure whether they were mobile.) Ancient extinct mayflies, most palaeodictyopteridans, and both ancient and modern odonatopterans all have two pairs of very similar or "homonomous" wings. Many neopterans,* in contrast, now have one pair of wings dominant and the other pair reduced (flies, for instance), although some of the less specialized neopterans still have two pairs of large, independently flapped wings.

Grasshoppers, for instance, are fairly unspecialized neopterans with large front wings and hindwings. They flap the front wings and hindwings out of phase, with the hindwings about a quarter stroke earlier than the forewings. In other words, when the forewings are just starting their downstroke, the hindwings are already about one-fourth of the way into their downstroke. We have good evidence that wing muscles evolved from leg muscles in insects—some modern insects have a pair or two of muscles that they use for both walking and flying—and the three pairs of legs also step out of phase during walking to avoid interfering with each other. So the grasshopper flapping pattern may be close to the original primitive pattern.

Dragonflies and damselflies have taken this pattern and amplified it. In the simpler grasshopper pattern, the front wings and hindwings always stay about one-quarter of a stroke out of phase, and the forewing makes most of the stroke adjustments the grasshopper needs for steering or acceleration. Not so in dragonflies. Dragonflies can shift from their more common out-of-phase pattern to the less common simultaneous or in-phase stroke pattern. In the out-of-phase pattern, when the forewings start their downstroke, the hindwings are just about finished with their

---

* Members of Neoptera, with the neopteran (folding) wing hinge; most flying insects are in this lineage.

downstroke. Moreover, forewings and hindwings are separately adjustable so that a dragonfly can adjust only the forewings, only the hindwings, or both for maneuvers like steering, climbing, or diving.[42] This great increase in adjustability of their wingbeat pattern clearly contributes to the spectacular maneuverability displayed by flying dragonflies. As a rule, when aerial predators are much larger than their prey, their higher flight speeds prevent them from turning as tightly as their prey. Dragonflies' extreme maneuverability seems to at least partly offset this size problem because dragonflies are highly successful aerial predators of tiny flying insects.

This adjustability comes with a cost: the wing articulations of the forewing and hindwing are basically right next to each other, and when the forewings are doing something very different from the hindwings, their sheer physical proximity means they tend to interfere with each other. Dragonflies avoid at least a little of this cost by having separate forewing and hindwing muscles attached directly to the wing articulations; most other insects have "indirect" flight muscles that actually deform the thorax to move the wings rather than pulling directly on each wing, and this reinforces the mechanical linkage of the front and back wings. To make matters worse, the wake of the air flowing over the front wing can actually interfere with lift production of the hindwing when it passes through the wake (which is why biplane airplanes have wings stacked one above the other rather than one on the same level close behind the other). Obviously, dragonflies have evolved the necessary strength and coordination to overcome this interference.

Most neopterans have evolved in a different direction where one pair of wings becomes dominant and the other pair is either reduced or takes on a new function (as we will see shortly). The vast majority of insects are neopterans, and they include the insects most familiar to nonspecialists, such as grasshoppers and crickets, praying mantises, cockroaches, termites, aphids, cicadas, bees, ants, wasps, beetles, butterflies, mosquitoes, and house flies. The neopteran wing hinge allows insects to flex their wings back so that the wings lie flat over the abdomen. Being able to stow the wings out of the way means these insects can live in nooks, crannies, and burrows that would be too cramped to inhabit with wings held out to the side like a dragonfly or held straight up over the back like a mayfly. We must be careful not to confuse correlation with causation, but logic dictates that neopterans should be capable of moving easily in spaces closed to more primitive flying insects, and neopterans are far and away the most diverse and abundant insects. Indeed, scientists have described more species of neopterans than all other species of animals combined. Ecologically and evolutionarily, Neoptera includes the most successful animals on the planet.

Flying insects have evolved along a fundamentally different trajectory than flying vertebrates like birds or bats. First, insects did not sacrifice any legs to gain wings, so insects have not given up any terrestrial mobility to achieve flight. Second, the original insect pattern is based on two pairs of wings rather than the single pair used by pterosaurs, living birds,* and bats. And finally, insects do not develop large, movable wings until they are fully adult, that is, full-grown and sexually mature. Birds and bats possess semi-functional wings as juveniles,[43] but the young grow very rapidly and are near adult size when they begin to fly competently, even if they are not sexually mature. Pterosaurs apparently grew more slowly and seem to have had functional wings over a wide range of body sizes (Chapter 8) although we have little evidence of their flight capabilities and not all researchers are convinced that the young ones could actually fly.

## JUVENILES WITH WINGS?

Thanks to that frustrating gap in the insect fossil record, we have no direct evidence about whether the earliest insects to achieve flapping flight had functional wings before adulthood. The indirect evidence is, unfortunately, somewhat conflicting. I have already described how mayflies, the living representatives of the most ancient known flying insects, have a pre-adult, fully winged stage; this could suggest that the earliest flying insects had fully winged juveniles. On the other hand, the wings of all other flying insects are physiologically and anatomically incapable of being molted due to that missing epidermis in the wing membranes. And since molting is a fundamental part of growth and development in insects, if the earliest flying insects had wings like most living insects, they could not have had functional wings before the adult stage. Moreover, at least one scientist thinks that the mayfly subimago may actually be a specialized adaptation to ease the mayfly's transition between an aquatic stage and an aerial stage rather than a primitive holdover from non-flying ancestors.[44]

---

* Paleontologists have now discovered several species of apparently four-winged birdlike dinosaurs (see Chapter 6), so birds may have also started out with two pairs of wings. Unlike insects, however, these four wings incorporated the animal's entire inventory of legs.

The nymphs of modern insects with incomplete metamorphosis have wing pads in their later stages; these are tiny winglets immovably glued down onto the thorax and completely nonfunctional. Some researchers have pointed out that fossils of nymphs of ancient, extinct mayflies have much larger wing pads, perhaps large enough to have been used for swimming or perhaps even very rudimentary flapping (see Fig. 5.7).[45] Some have even argued that these nymphs are evidence for some variation on the gill/exite theory of wing evolution. A closer look suggests otherwise.

First, the fossils do not appear to have a well-developed, flexible articulation. Second, the wing pads are bent sharply backward, as is obvious in Figure 5.7. They are not flexed backward at the hinge, as in a modern house

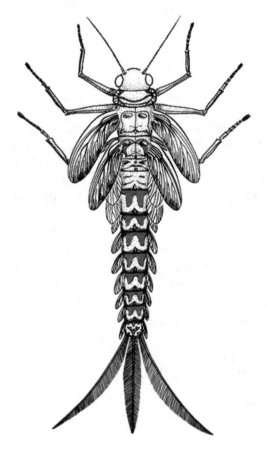

**Figure 5.7:**
Reconstruction of fossil mayfly nymph, from Kukalova-Peck, used by permission of John Wiley & Sons.[32]

fly; instead the whole wing pad itself is curved backward. In this orientation, the wing pad would be quite useless as either a wing or a swimming paddle. These wing pads clearly were not being used for locomotion. Most likely, these are fossils of older nymphs nearly ready to molt to adults, and these primitive insects had larger wing pads because the developmental programming to allow more dramatic wing size changes between molts had not yet evolved. In other words, large wing pads in juveniles might have been a developmental necessity early on, and later insects then evolved more sophisticated development so nymphs did not need to carry around large, nonfunctional wing pads just so adults could have big wings.

**FOUR WINGS VERSUS TWO**

Although the primitive insect pattern is based on two pairs of wings and we have seen how dragonflies have made a virtue of that necessity, more recently evolved insect groups have repeatedly evolved a single dominant pair of wings that do most of the aerodynamic work. We can't go back in time and do controlled experiments on evolution so we cannot clearly demonstrate cause-and-effect, but we do know that independently flapping fore- and hindwings can interfere with each other aerodynamically and that the articulations of fore- and hindwings flapping out of phase can interfere with each other mechanically. In the majority of flying insects, the forewings are dominant and the hindwings are reduced or vestigial. In beetles and earwigs, however, the forewings have evolved into hard, protective wing covers too small to fly with, so the hindwings are the main flight surfaces and must be folded origami-style to fit under the much smaller wing covers. In other insects, where the forewings rather than hindwings are dominant, wing muscles hold the front of the smaller hindwing in contact with the back of the forewing; the fore- and hindwings are flapped simultaneously, that is, in phase. They function, in fact, like a single pair of wings. This pattern is used by butterflies and a few moths. Many such insects have gone even further and evolved attachment or locking mechanisms: hooks or bristles or Velcro-like structures to hold the hindwings to the front wings to form a single surface. Bees and wasps, aphids, cicadas, and stink bugs are examples of insects that use locking mechanisms to hold front and back wings together while flapping.

Some insects have carried dominance of the forewings to an extreme. Dipterans or true flies—including house flies, horse flies, blow flies, bee flies, mosquitoes, and midges—appear to have but a single pair of wings, corresponding to the forewings of other insects. Upon close inspection, we can see that flies have a tiny club-shaped appendage behind each wing, not much bigger than a large bristle. These minute structures, called "halteres," are actually hindwings that have evolved into sensory structures. The halteres swing up and down when the wings are flapping and have properties similar to a gyroscope. They can detect unintentional heading changes, as when the fly is deflected by a gust, and they control a set of reflexes that adjusts the wing stroke pattern to correct any unintended course deviations.[46]

## LOSS OF FLIGHT

Scientists have named approximately 1 million species of insects, and estimates of the number of those yet to be named range from 2 million to 30 million. The overwhelming majority of insects can fly; yet given this vast diversity, we should not be surprised if here and there across the insect phylogeny we find scattered pockets of insects that have abandoned flight in their evolution. In fact, possibly as many as 10% of the members of the flying insect lineage have lost the ability to fly.

Some insects, such as silverfish, have no wings because they branched off the main lineage of insects before wings evolved. Other insects, fleas for example, evolved from winged ancestors. We have seen evidence that tells us the silverfish are primitively wingless, but how can we tell that fleas and many other insects are secondarily wingless, that is, descended from winged ancestors? Some are easy: fleas have complete metamorphosis (distinct larval, pupal, and adult stages) and scientists have abundant evidence that complete metamorphosis evolved long after insects evolved flight. For others, phylogenetic analysis leaves little doubt. Lice, which are all wingless, have incomplete metamorphosis, but unlike silverfish, they never molt as adults. Moreover, lice are closely related to the neopteran order Psocoptera (book lice, most of which are winged), and so they must clearly be neopteran. Or consider bedbugs. Bedbugs are in the family Cimicidae, a family of a few dozen blood-feeding species, only two of which prefer humans. The entire family is wingless, but they are clearly nested within the order Hemiptera or true bugs, most of which are winged. Bedbugs' closest relatives can fly, so bedbugs clearly evolved from flying bugs.

Given the obvious advantages of flight, why have some insects lost this ability? Many factors could be involved, and scientists have begun to tease out some significant patterns. We will take a more detailed look at loss of flight in insects, as well as other groups, in Chapter 9. These secondarily flightless insects, however, are little more than a drop in the bucket compared to the overwhelming majority of insect species that can fly.

**REFERENCES**

1. P. Whalley and E. A. Jarzembowski (1981) *Nature*.
2. W. A. Shear, P. M. Bonamo, J. D. Grierson, et al. (1984) *Science*.
3. M. S. Engel and D. A. Grimaldi (2004) *Nature*.
4. D. A. Grimaldi and M. S. Engel (2005) *Evolution of the Insects*.
5. A. H. Staniczek, P. Sroka, and G. Bechly (2014) *Systematic Entomology*.
6. C. Gegenbaur (1870) *Grundzüge der vergleichenden Anatomie*.
7. C. Gegenbaur (1878) *Elements of Comparative Anatomy*.
8. F. Müller (1873) *Jenaische Zeitschrift für Naturwissenschaft*.
9. F. Müller (1875) *Jenaische Zeitschrift für Naturwissenschaft*.
10. R. Dudley (2000) *The Biomechanics of Insect Flight: Form, Function, Evolution*.
11. W. T. M. Forbes (1943) *American Midland Naturalist*.
12. R. D. Alexander and W. L. Brown (1963) *Occasional Papers of the Museum of Zoology of the University of Michigan*.
13. V. B. Wigglesworth (1973) *Nature*.
14. O. Béthoux and D. E. G. Briggs (2008) *Systematic Entomology*.
15. D. A. Grimaldi (2010) *Arthropod Structure & Development*.
16. M. M. Douglas (1981) *Science*.
17. J. W. Flower (1964) *Journal of Insect Physiology*.
18. J. G. Kingsolver and M. A. R. Koehl (1985) *Evolution*.
19. J. G. Kingsolver and M. A. R. Koehl (1994) *Annual Review of Entomology*.
20. R. J. Wootton and C. P. Ellington (1991) in *Biomechanics in Evolution*.
21. J. H. Marden and M. G. Kramer (1994) *Science*.
22. J. H. Marden and M. G. Kramer (1995) *Nature*.
23. J. H. Marden (2003) *Acta Zoologica Cracoviensia*.
24. M. S. Engel, S. R. Davis, and J. Prokop (2013) in *Arthropod Biology and Evolution: Molecules, Development, Morphology*.
25. S. P. Yanoviak and R. Dudley (2006) *Journal of Experimental Biology*.
26. S. P. Yanoviak, R. Dudley, and M. Kaspari (2005) *Nature*.
27. S. P. Yanoviak, Y. Munk, M. Kaspari, et al. (2010) *Proceedings of the Royal Society Biological Sciences Series B*.
28. I. Hasenfuss (2002) *Journal of Zoological Systematics and Evolutionary Research*.
29. A. Jusufi, Y. Zeng, R. J. Full, et al. (2011) *Integrative and Comparative Biology*.
30. V. B. Wigglesworth (1963) *Nature*.
31. V. B. Wigglesworth (1976) in *Insect Flight*.
32. R. Garrouste, G. Clement, P. Nel, et al. (2012) *Nature*.
33. T. Hornschemeyer, J. T. Haug, O. Béthoux, et al. (2013) *Nature*.
34. J. Kukalova-Peck (1978) *Journal of Morphology*.
35. R. J. Wootton (1972) *Palaeontology*.
36. F. M. Carpenter (1992) in *Treatise on Invertebrate Paleontology, Part R, Arthopoda 3–4*.
37. A. J. Ross (2010) *Scottish Journal of Geology*.
38. W. Hennig (1981) *Insect Phylogeny*.

39. J. Kukalova-Peck (1997) in *Arthropod Relationships*.
40. J. Kukalova-Peck (1985) *Canadian Journal of Zoology*.
41. R. J. Beckemeyer (2000) *Kansas School Naturalist*.
42. D. E. Alexander (1986) *Journal of Experimental Biology*.
43. K. P. Dial (2003) *Science*.
44. V. C. Maiorana (1979) *Biological Journal of the Linnean Society*.
45. J. Kukalova-Peck (1983) *Canadian Journal of Zoology*.
46. D. E. Alexander (2002) *Nature's Flyers: Birds, Insects, and the Biomechanics of Flight*.

## CHAPTER 6
# Birds

*The Feathered Flyers*

Say "flying animal" and most people will probably think "bird." Birds are certainly the most conspicuous and probably the most familiar of flying animals. Birds communicate with sounds, like us, and they are warm-blooded, like us, and many familiar birds (especially males) have showy, visually striking plumage. Humans often seem to feel a stronger affinity to birds than to other mammals, even though we are much more distantly related to the birds. Birdwatching is both a popular individual hobby and the focus of numerous regional and national organizations, but I have yet to hear of anyone going "rodent-watching" or joining a "mammal-watching" society. This affinity for birds may explain why more effort has gone into studying the evolution of bird flight than the flight of all other animals combined.

**FIRST FOSSILS**

Any discussion of the fossil history of birds logically begins with *Archaeopteryx*, the classic "missing link" fossil between birds and dinosaurs. (Indeed, researchers who study insect or bat evolution often lament the lack of an *Archaeopteryx*-like fossil for those groups.)[1,2] Several magnificent specimens of *Archaeopteryx* were found in the Solnhofen limestone of Germany as it was being quarried for use in making lithographic prints. The smooth, fine-grained texture of Solnhofen limestone that makes it so highly prized for lithography also makes it an excellent medium for preserving impressions of delicate soft tissue structures of entombed organisms. Two of the

**Figure 6.1:**
The "Berlin" *Archaeopteryx* fossil, housed in the Humboldt Museum für Naturkunde in Berlin. These animals are smaller than most people expect; this specimen was just slightly bigger than a pigeon, with a body length of about 23 centimeters (9 inches) not counting the long tail. (Photo courtesy of David A. Burnham, used by permission.)

original *Archaeopteryx* specimens, the "London" specimen discovered in 1861 and the "Berlin" specimen found in 1877, clearly preserved extensive birdlike plumage, including unmistakably aerodynamic wing feathers (see Fig. 6.1).[3,4] Yet the body was clearly reptilian, with teeth, long separate fingers with claws, and a long, whip-like bony tail (albeit covered with feathers). Coming hard on the heels of Darwin's publication of his theory of evolution in 1860,[5] these fossils must have seemed tailor-made to support his new concept. Over the years, several more specimens of *Archaeopteryx* have been found, and although most have feather impressions, none show such clear and detailed feathers as those first two.[6]

For nearly a century and a half, biologists and paleontologists accepted *Archaeopteryx* as the oldest and most primitive bird known, having lived about 160 million years ago. Its feathers were so birdlike and its wings were so clearly aerodynamic that to suggest that *Archaeopteryx* was not at the base of the bird family tree was to invite ridicule. Scientists generally agreed that if *Archaeopteryx* itself was not ancestral to modern birds, it was certainly a close relative to that ancestor. In the past, some scientists have questioned whether *Archaeopteryx* was a powered flyer or whether it could even glide,[7] although most now view it as at least a weak powered flyer. Recent fossil finds in China have complicated our view of the relationships among ancient birds and other dinosaurs, sometimes displacing *Archaeopteryx*, as we'll see in this chapter. The iconic status of *Archaeopteryx*, however, makes researchers who move it anywhere outside the bird lineage on a phylogeny very careful to fully justify that placement.

## DESCENDED FROM DINOSAURS

Almost from the beginning of the modern scientific study of evolution, scientists linked birds with dinosaurs. After the spectacular London and Berlin specimens of *Archaeopteryx* were discovered and widely publicized, scientists quickly noticed that this species had birdlike feathers on a body with many dinosaurian features. For decades, most scientists took *Archaeopteryx* as clear evidence that birds evolved from dinosaurs. In the 1920s, this belief took a curious detour. Danish artist Gerhard Heilmann wrote some articles on bird evolution that were published in book form (in Danish) in 1916 and largely ignored. In 1926, he published an English translation, *The Origin of Birds*, which became extremely influential. In it, he said that, yes, *Archaeopteryx* does seem to have a lot of dinosaur-like characteristics, but birds all have a furcula ("wishbone") formed from modified clavicles or collarbones, and dinosaurs have no clavicles. He thus reluctantly concluded that birds could not have descended directly from dinosaurs but must instead share with them some more-ancient common ancestor.[8] Heilmann seems to have stated his view so carefully and forcefully that "birds descended from dinosaurs" quickly became a minority view.

Heilmann's book prompted researchers to propose possible ancestors for birds other than dinosaurs. Usually these were "thecodonts," a collection of ancient, primitive, dinosaur-like reptiles. At the time, paleontologists considered thecodonts to be a group of closely related animals ancestral to dinosaurs, although we now view them as a miscellaneous collection of not-so-closely-related animals, of which one or two may be related to dinosaur ancestors. Since *Archaeopteryx* was so much like a dinosaur, looking for the common ancestor of birds and dinosaurs among thecodonts largely amounted to looking for the closest relatives of dinosaurs in that group.

Other researchers proposed that birds arose from the crocodilian lineage. Modern crocodiles and their relatives might seem an unlikely source for bird ancestors, but proponents made reasonable arguments. First, crocodiles, birds, dinosaurs, and pterosaurs make up the lineage Archosauria, and at the time almost nothing definitive was known about who was related to whom among the four archosaurian groups. Moreover, not all of the early crocodilians were the long, sprawling, semi-aquatic beasts of today. The most ancient crocodilians were much more diverse and included bipedal runners that some scientists thought may even have been able to climb trees.[9] *Archaeopteryx* teeth also seemed to be more crocodilian than dinosaurian, at least based on the dinosaurs known in the mid-20th century. Although never a majority view, the crocodilian-bird-ancestry theory was advocated by some respected scientists.[10]

In the 1970s and 1980s, several events shifted views of bird ancestry back toward dinosaurs. First, paleontologists started using modern phylogenetic methods (Chapter 1), which uniformly showed a close relationship among dinosaurs, *Archaeopteryx*, and birds. Second, eminent Yale paleontologist John Ostrom had his epiphany linking *Archaeopteryx* to small theropod (meat-eating) dinosaurs (we will look in detail at Ostrom's important insight later). And, not least, researchers found clavicles (and even furculas) on more complete specimens of dinosaur fossils.[11,12] The evidence has become quite strong that birds descended from theropod dinosaurs. Specifically, birds descended from a diverse lineage of theropod dinosaurs called Maniraptora. Figure 6.2 illustrates how Maniraptora and other groups important to bird evolution form subgroups within dinosaurs. Maniraptora includes predatory dinosaurs like the mollusk-eating *Oviraptor* and the weird, short-armed *Mononykus*, as well as those with the enlarged, retractable toe-claw such as *Velociraptor* and *Deinonychus* (see Box 6.1).

Scientists argue about whether *Archaeopteryx* should be considered a bird or a birdlike dinosaur, but most recent phylogenies place *Archaeopteryx* on a side branch off the very base of the bird lineage (see Fig. 6.2), so we will call *Archaeopteryx* a basal bird. Nowadays, paleontologists call this bird lineage of the maniraptoran family tree "Avialae" ("bird wing"). It was originally defined as all feathered animals capable of flapping flight[13] so as to distinguish it from Aves, the traditional classification for the living birds. (Avialae thus includes all living birds and extinct birds back to *Archaeopteryx*—and arguably a couple of species even older than

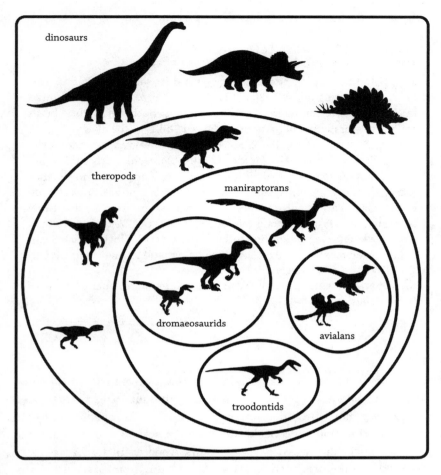
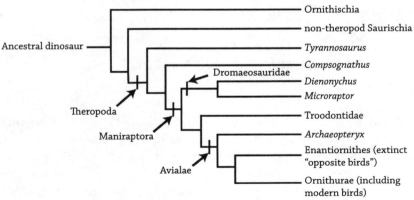

**Figure 6.2:**
(Top) This Venn diagram names many of the dinosaur subgroups close to the bird lineage and shows how they are nested. (Bottom) A phylogeny demonstrating the same relationships. (Courtesy of S. T.)

> *Box 6.1:* THE "REAL" *VELOCIRAPTOR*
>
> Scientists know of *Velociraptor* from several fossils found in and near Mongolia starting in the 1920s. It is very similar to *Deinonychus*, found in North America and described in the 1960s. Both were lightly built, agile predators with the famous, enlarged "killing" toe claw; *Deinonychus* actually means "terrible claw." (Scientists disagree on whether the enlarged claw was actually for killing prey; biomechanical studies suggest it may have been used to hold down smaller prey or in defense rather than in slashing attacks.) The public knows of *Velociraptor* from the "Jurassic Park" book and movies, which, however, were not entirely accurate. At less than 1 meter (about 2 or 3 feet) tall, *Velociraptor* would not have been a very menacing villain. The movie "velociraptors" seem to have been mostly based on the much larger (and scarier) *Deinonychus*. Thanks to the movies' terminology, however, "velociraptor" has entered the general lexicon whereas few people other than scientists know of *Deinonychus*.

*Archaeopteryx*.) Since Avialae is a lineage that evolved within theropod dinosaurs, then phylogenetically birds are dinosaurs; this is the reason we must say "non-bird dinosaurs" or "non-avialan theropods" to refer to groups we traditionally consider dinosaurs. For the same reason, when paleontologists are asked why dinosaurs went extinct, they tend to answer that dinosaurs are not extinct: today's dinosaurs are just a profusion of small, feathered animals we know as birds.

## THE ORIGIN OF FLAPPING FLIGHT IN BIRDS
### Two Traditional Theories

In an odd parallel to the situation in insects, scientists have been arguing for more than a century over two theories that explain how bird flight evolved. One theory starts with an arboreal (tree-climbing) protobird that then evolves from leaping, through a gliding stage, then through steering and weak flapping, to finally achieving fully powered flight. This is essentially the same theory that Darwin originally proposed for the evolution of flight in bats, and scientists call it the "arboreal" or "trees-down" theory. The other theory starts with a fast, bipedal runner. This runner evolves aerodynamic surfaces on its arms, perhaps to aid in balance or to stabilize turns while running. These surfaces enlarge and allow a bit of gliding during leaps. Flapping would have evolved initially for thrust only—to increase running speed. Flapping in time became strong enough to turn leaps into short flights and eventually evolved into fully powered flight. Scientists call this the "cursorial" or "ground-up" theory. Some proponents of the cursorial theory see no need for a gliding stage at all: they see flapping for thrust (to increase running speed) as sufficiently beneficial by itself that flapping could have evolved directly. In that form, the cursorial theory achieves powered flight in almost the same way as Marden's theory of surface-skimming stoneflies (Chapter 5).

One problem with the arguments over these theories is that researchers kept trying to use evidence from *Archaeopteryx* to support one or the other. Supporters of the cursorial theory would say, "*Archaeopteryx* has features *a* and *b*, making it a runner, so flight must have evolved from the ground up." Then supporters of the arboreal theory would say, "*Archaeopteryx* has features *x* and *y*, making it a climber, so flight must have evolved from the trees down." Both sides missed the point that *Archaeopteryx* was past the evolutionary stage where being arboreal or cursorial mattered. *Archaeopteryx* wasn't in the process of evolving flight, it was already a flyer. It may not have been the most powerful, agile flyer, but its wing shape and wing loading are well within the range of modern birds,* and it appears to have been capable of at least some powered flight.[14] The bottom line is that *Archaeopteryx* is very important for unraveling the phylogenetic relationships of birds and their relatives (which can help researchers figure out which characteristics were inherited from ancestors), but its flight abilities are advanced enough that it may not tell us much about the beginnings of flight in birds.

## A Misleading Linkage

Another problem was with researchers' tendency to mix up ancestry with flight evolution. Yale paleontologist John Ostrom made discoveries that clarified the phylogenetic position of *Archaeopteryx* but that also misdirected flight evolution research. Ostrom had discovered and described *Deinonychus*, so he was intimately familiar with the dinosaurs that would later be called maniraptorans. He then turned his attention to pterosaurs.

---

* Wing loading is the lift per unit surface area of a wing and relates wing size to body weight.

He traveled to the Netherlands to visit the Teyler Museum to see a pterosaur fossil that had been described a few years before *Archaeopteryx* in the 1800s. As soon as he looked at it, he knew the fossil was not a pterosaur, and on closer inspection he discovered faint feather impressions: it was actually an unrecognized fossil of *Archaeopteryx*![6] Ostrom shifted his research focus toward *Archaeopteryx*, and the more he looked at *Archaeopteryx* fossils, the more he saw similarities to *Deinonychus*. In the 1970s, Ostrom pointed out these similarities,[15,16] and other paleontologists began building phylogenies that clearly placed *Archaeopteryx* in or near the *Deinonychus* lineage.[13] To these paleontologists, *Archaeopteryx*—and by extension all birds—had now become a subgroup of dinosaurs.

The misdirection came from combining the new phylogenies with a view of *Archaeopteryx* as if it were just beginning to evolve flight and an incomplete knowledge of Maniraptora. These factors led most paleontologists to argue that *Archaeopteryx* is clearly a close relative of theropod dinosaurs, and since theropod dinosaurs are bipedal runners and not tree climbers, flight must have evolved from the ground up.

For several decades before Ostrom looked at *Archaeopteryx*, the arboreal theory seems to have been more widely accepted.[17,18] After Ostrom revived the cursorial view, two groups of scientists remained unconvinced. One group included many ornithologists, who considered *Archaeopteryx* to be a bird and who were skeptical of claims of close kinship between *Archaeopteryx* and theropod dinosaurs.[19] The opinion of these researchers was essentially a mirror image of the view put forth by Ostrom and the dinosaur paleontologists: birds evolved from some lineage much more ancient than Maniraptora; *Archaeopteryx* was not a runner so it could not have descended from theropods; and since bird ancestors were climbers (presumably arboreal), flight must have evolved from the trees down.

The arguments between the cursorial supporters and the arboreal supporters,* although couched in dispassionate scientific terms, tended to sound a bit like "whatever they are for, I'm against." In hindsight, the "not from theropods/trees down" proponents were also misdirected. Because they favored an arboreal origin for flight (and they viewed *Archaeopteryx* as a bird and not a dinosaur), they felt compelled to oppose a close theropod-bird relationship.

---

* In addition to Ostrom, prominent supporters of the "theropod ancestry/ground up" position include UC-Berkeley paleontologist Kevin Padian and his former student, Jaques Gautier. On the other side, University of North Carolina ornithologist Alan Feduccia and my University of Kansas colleague, the late Larry Martin, championed the "more ancient ancestry/trees down" view.

## Does Physics Matter?

Yet a third group of researchers was drawn into this debate. These were the flight biomechanists, scientists studying the aerodynamics and functional anatomy of bird flight (full disclosure: my own research falls closest to this group). Prominent members of this group included British biologist Jeremy Rayner and Swedish biologist Ulla Norberg. These researchers focused on the physics of flight and were much less concerned about phylogenetic relationships. They pointed out (repeatedly) that from a physical standpoint, the arboreal model makes much more sense than the cursorial model.[14,20-23] The fundamental point is that if you start out gliding from an elevated perch, you are working with gravity and you gain all the advantages of gliding (Chapter 4). In contrast, Rayner analyzed the physics and showed that *Archaeopteryx*-sized animals cannot run fast enough to gain any benefit from leaping and gliding.[24]

Proponents of the ground-up model responded that flapping could have evolved directly, skipping a gliding stage, to produce thrust for faster running.[25] Using flapping wings to run faster is certainly possible—chickens do it all the time—but evolving wings just for that purpose seems unlikely. First, protowings would have to get fairly big and flapping muscles would have to get pretty strong before they could produce enough extra thrust to help creatures run faster. This is an especially large hurdle to overcome because simply evolving longer legs can do the same thing. Second, not a single animal, living or extinct, is known that uses flapping appendages to run faster on the ground *except* those descended from flapping flyers. Indeed, flightless cursorial birds like ostriches and emus are excellent runners and are even descended from flapping flyers, yet they do not use their wings to increase running speed; their wings are essentially vestigial. In the early 1990s, these three groups of scientists seemed to be at an impasse. No one could marshal enough evidence to sway members of the other groups. Then the first of what became a flood of new fossils appeared. These new fossils have forced researchers to re-evaluate their arguments.

## EXPLOSION OF FOSSILS FROM CHINA

The fossil beds in Liaoning Province of northern China have produced a wealth of fossil mammals, insects, plants, extinct birds, and pterosaurs from the early Cretaceous, about 150 million years ago. They are best known to both scientists and the general public, however, for the abundant dinosaur fossils they have produced.

As political conditions for scientific research and collaboration in China improved in the 1980s and 1990s, paleontologists began describing a variety of very well-preserved dinosaurs from the Liaoning beds. Among these were theropod dinosaurs that showed a close relationship to *Deinonychus* and *Velociraptor* (including the famous enlarged toe-claw), placing them in the Maniraptora. These dinosaurs gave researchers more data to refine their phylogenetic trees, and the more detailed the trees became, the stronger the evidence looked for placing *Archaeopteryx* in that group as well.

Like the Solnhofen limestone, the fine-grained texture of the rock of these Chinese fossil beds often preserves extremely fine detail, including what appear to be hair-like or feathery external structures. Some of the first fossils to show such structures were specimens of *Confuciusornis sanctus*, a very primitive beaked bird. These sparrow-sized birds are extremely common among Liaoning fossils, often including preserved feathers. They are remarkable because a few have a pair of very long tail feathers whereas most have only very short tail feathers. Although the feathers were preserved in very fine detail, no one was surprised by the presence of feathers; these were birds, after all.

### Feathered Dinosaurs?

In 1998, a team of Chinese and North American scientists reported a startling find: two different dinosaur fossils that both showed what appeared to be feathers.[26] *Protarchaeopteryx* and *Caudipteryx* were both small (goose-sized) maniraptoran theropod dinosaurs.* Both fossils had clear impressions of what appeared to be different types of feathers: short, fluffy, filamentous or down-like feathers and longer "pennaceous" feathers. Pennaceous feathers are what we usually envision when we think of feathers: a central shaft with a flat array of interlocking barbs on each side forming a vane or blade. *Protarchaeopteryx* had a fringe of fluffy feathers along its chest, legs, and tail, plus several pennaceous feathers on its tail. The *Caudipteryx* fossil, on the other hand, showed a well-developed row of obvious pennaceous feathers on each hand and on its tail (in addition to a fringe of downy feathers surrounding its body). Other than their small size—and feathers—these dinosaurs were fairly typical bipedal, carnivorous, maniraptoran dinosaurs. We'll see that their

---

* Recent phylogenies place them in the ovoraptorids, one of several subgroups of Maniraptora.

feathers, though pennaceous, did not have the right shape or arrangement to form a wing, so they certainly could not fly (or even glide effectively). Aside from their feathers, they were no more birdlike than many other maniraptoran dinosaurs.

When earlier researchers had described hair-like fibers fringing the skeleton of a Liaoning dinosaur as protofeathers,[27] other scientists dismissed them as tendons or skin fibers that only appeared to be external due to fossilization processes.[28,29] With the discovery of unmistakable pennaceous feathers on *Caudipteryx*, a close relationship between maniraptoran dinosaurs and birds became unmistakable. Researchers developed phylogenies showing relationships among maniraptoran dinosaurs, *Archaeopteryx*, and other birds, and these required constant updating as new fossil discoveries from Liaoning steadily increased. The paleontologists who described *Caudipteryx* nearly two decades ago would probably be amazed at the huge increase in known maniraptoran dinosaur species as well as the number of those dinosaur species that show signs of feathers. For example, with only minimal effort, I tracked down information on over two dozen non-bird, maniraptoran dinosaurs described after *Caudipteryx*. Guided by a recent phylogenetic analysis of over 100 avialans (bird lineage animals) and non-avialan theropods,[30] I counted at least 10 with apparent down-like feathers and at least 10 more with unmistakable pennaceous feathers. We now know of a number of theropods that had downy filaments and are only distantly related to maniraptorans, including two species closely related to *Tyrannosaurus rex*!

### Four Wings?

Among the many startling fossils from Liaonning, perhaps the most remarkable is *Microraptor gui* (Fig. 6.3). *Microraptor* was a small (1-kilogram or 2-pound) maniraptoran dinosaur with long, asymmetrical pennaceous flight feathers on both its arms and legs. This peculiar beast appears to have been a four-winged flyer. Modern flying birds have asymmetrical primary (pinion) feathers making up the wingtip, with the shaft closer to the front edge of the blade rather than in the center (Fig. 6.4). In flight, this asymmetry stabilizes the wingtip and also lets each primary feather act as a tiny separate wing when the primaries are spread out during parts of the wing stroke.[31] In contrast, *Caudipteryx*'s arm feathers were symmetrical, so they certainly were not acting as pinion feathers (Fig. 6.4). The asymmetrical feathers of *Microraptor*, however, are convincing evidence that both its arms *and* its legs functioned as wings. Although possessing four

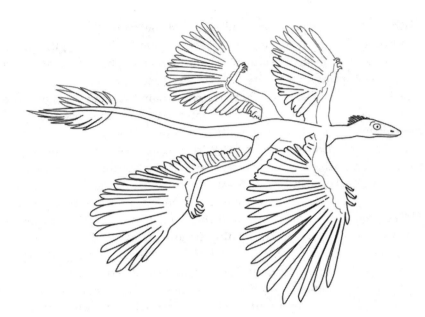

**Figure 6.3:**
*Microraptor gui* reconstructed as a four-winged glider. (Courtesy of S. T.)

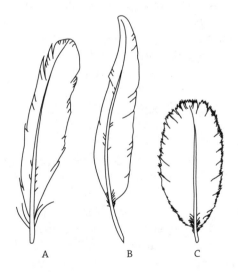

**Figure 6.4:**
Primary (pinion) feathers from different birds. A. *Archaeopteryx*. B. Modern flying bird (pigeon). C. Emu, a flightless bird. The primary feathers from flying birds are always asymmetrical, with the central shaft of the feather closer to the front of the wing. When birds lose the ability to fly, the feathers tend to become more symmetrical, like the secondary feathers on the back of the wing and the contour feathers that cover the body. (From Alexander.)[29]

well-developed wings, *Microraptor* was apparently a glider and not a fully powered flyer. It did not have the shoulder joint or chest muscles for well-developed flapping, although it may have used very weak flapping to extend glides.

*Microraptor* is significant in several ways. It is clearly a maniraptoran dinosaur, and it is just as clearly arboreal. In fact, the feathers on its feet were about 18 centimeters (7 inches) long, on a leg of only 26 or 28 centimeters (10 or 11 inches) from hip to toes, so I have a hard time imagining how it could walk on the ground at all; it was clearly not a bipedal runner. The existence of *Microraptor* means that the proponents of the cursorial model of flight evolution can no longer say "birds evolved from theropods, and all theropods are bipedal runners, so birds must have evolved flight from the ground up." *Microraptor* provides solid evidence that theropods could be arboreal and could evolve gliding.

*Microraptor* lived a few million years after *Archaeopteryx*, so cannot be directly ancestral to birds. Nevertheless, *Microraptor* shows that theropods could evolve an arboreal lifestyle and could evolve gliding from the trees down. Some scientists consider *Microraptor*'s front wings to be so birdlike that birds and *Microraptor* must have inherited these wings from a common ancestor. Others doubt that birds and *Microraptor* share a common gliding ancestor, but they see *Microraptor* as establishing a possibility: the possibility that a dinosaur similar to *Microraptor* evolved gliding much earlier, and that gliding dinosaur gave rise to *Archaeopteryx* and the rest of the bird lineage.

One recent phylogenetic study actually supports a gliding, four-winged common ancestor for *Microraptor* and birds.[30] Figure 6.5 shows a simplified

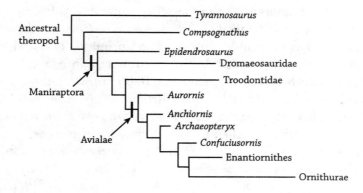

**Figure 6.5:**
Part of the phylogeny of Godefroit et al., redrawn and with some groups omitted for clarity.[30] All these groups are extinct except for one lineage of Ornithurae leading to modern birds.

version of their phylogeny. Many of the more primitive members of the Avialae, including *Anchiornis, Archaeopteryx, Sapeornis, Confuciusornis,* and some enantiornithines, had long leg feathers, again demonstrating a link to a *Microraptor*-like body plan.[32]

## ARCHAEOPTERYX DETHRONED

Quite a number of feathered theropod dinosaurs are now known, including both distant and close relatives of avialans. Paleontologists have described several feathered maniraptorans in recent years that would unquestionably have been considered typical dinosaurs if they had not possessed feathers. How do these feathered and birdlike dinosaurs affect our view of bird ancestry? These finds have forced scientists to keep modifying their phylogenies. Sometimes these modified phylogenies move *Archaeopteryx* out of Avialae[33] while other studies put it back into the avialan lineage,[34] usually at the base. Moving *Archaeopteryx* out of Avialae has profound implications: either flapping flight evolved twice (once in *Archaeopteryx* and once in Avialae), or all the maniraptoran dinosaurs close to Avialae (*Velociraptor, Deinonychus,* other dromaeosaurids and troodontids) are secondarily flightless. Most scientists would consider either of these possibilities as very unlikely (although at least a couple support the latter),[35] so the strong consensus is that flight evolved only once in the bird lineage, a view that keeps *Archaeopteryx* within Avialae.

Recent studies challenge the First Bird status of *Archaeopteryx*, however. Pascal Godefroit and colleagues published a description of another feathered maniraptoran, *Aurornis xui*. They combined this description with by far the most detailed phylogeny ever developed for theropods, including dozens of birdlike dinosaurs and primitive extinct birds—more than 100 total species.[30] In their most likely phylogeny, both *Aurornis* and another well-feathered maniraptoran, *Anchiornis huxleyi*, appeared at the base of Avialae, even further back in time than *Archaeopteryx* (Fig. 6.5 shows part of their phylogeny). Both these newcomers had long, flat feathers on their arms and legs, although the feathers are not as aerodynamically specialized as *Archaeopteryx* wing feathers,* so they were probably gliders or at best very weak flappers. *Anchiornis* was

---

* *Auronis* and *Anchiornis* wing feathers are symmetrical rather than asymmetrical like those of *Archaeopteryx*.

originally described as a troodontid, a group of large-brained, feathered maniraptorans generally considered to be the closest non-bird relatives of Avialae (Troodontidae, Fig. 6.5),[36] so even from the beginning, it was linked with birds. If Godefroit and company's analysis withstands the test of time, *Aurornis* would displace *Archaeopteryx* at the base of the lineage leading to birds. The large number of species and enormous number of anatomical features used in the study by Godefroit and colleagues give it good credibility, but only time will tell whether other scientists accept it and whether it remains consistent with future fossil discoveries.

Although a handful of scientists are still skeptical that the fluffy filaments on many Liaoning dinosaur fossils are really feathers,[37] the fossils showing pennaceous feathers seem to me to be unassailable evidence that a wide variety of non-bird theropods had feathers. The presence of unmistakable feathers on non-flying species like *Protarchaeopteryx*, *Caudipteryx*, and several other maniraptorans decisively answers one long-standing question in the evolution of bird flight: did feathers evolve specifically as an aerodynamic adaptation, or did they evolve initially for some other function and later get co-opted for flight? We now know that feathers evolved long before flight. Moreover, if those filamentous tufts on a wide diversity of theropods turn out to be protofeathers, that implies that most theropods may have been feathered. Theropods seem to have been active carnivores, and many, especially the later ones, may have been partly or completely endothermic (warm-blooded), so the filamentous protofeathers may have evolved as insulation to help retain heat.

The temperature-regulation question is still rather murky, however, because body size matters greatly, and theropods include dinosaurs with a huge range of body sizes, from the 6-ton *Tyrannosaurus* and 1½-ton *Gigantoraptor* to the 2- to 4 ½-kilogram (5- to 10-pound) *Bambiraptor* and ½-kilogram (1-pound) *Eosinopteryx*. We will return to this question when we look at physiological adaptations for flight. For now, suffice it to say that insulation can be very beneficial for small endothermic animals; as animals get big, however, its value decreases, and for really big animals like *Tyrannosaurus*, insulation has no physiological benefit at all. In other words, *Tyrannosaurus* would not have needed insulation to maintain a high body temperature. Yet two of its closest relatives, *Dilong* and *Yutyrannus* show traces of filamentous feathers, and the latter was also huge. If all these filaments on theropod dinosaurs really are protofeathers, either they have some function other than insulation or their function is different in really small and really big dinosaurs.

## NEW EVIDENCE, NEW VIEWS
### Pouncing Proavis

In 1999, British researchers Joseph Garner, Graham Taylor, and Adrian Thomas published one of the more innovative attempts to combine phylogenies and biomechanics. Their study set out to match the phylogeny of then-known feathered dinosaurs and birds with the order they expected to see various features evolve in either the arboreal or the cursorial models.[38] At the time, these scientists knew about *Protarchaeopteryx* and *Caudipteryx*, but discovery of small or feathered theropods like *Bambiraptor*, *Microraptor*, and *Eosinopteryx* lay in the future. The phylogeny of Garner and company led them to the conclusion that neither the arboreal nor the cursorial models exactly fit the order in which features appeared on their phylogenetic tree. For example, they described how *Caudipteryx*'s feathered hand fit better with a version of the cursorial theory, whereas *Archaeopteryx*'s long bony tail and dinosaurian hips fit better with the arboreal model.

They proposed instead that the ancestor of birds was a predator that specialized in pouncing on prey from elevated perches. If such predators mainly used their feet to attack prey (with that enlarged claw), their forelimbs could be used for balance and to help steer during the leap. At first, aerodynamic enhancement of the hands would have helped steer during leaps; then it could provide a bit of lift for "turning a pounce into a swoop,"[38] when lift production increased as a byproduct of more effective steering. *Caudipteryx* fits nicely into this scenario because the feathers on its hands are not well shaped or correctly oriented for flapping or gliding,* but they could be quite effective for steering in mid-air. Once these feathered "control surfaces" started to provide lift and extend leaps, selection for longer swoops would lead to more effective wings, then glides, and eventually flapping. They called their model the "pouncing proavis."

When I first read the article by Garner and colleagues, I thought, "At last, here is a fresh idea that can help bridge the gap between the arboreal and cursorial camps." In fact, the pouncing proavis theory has been largely ignored, and I was chagrinned to see essentially the same old cursorial and arboreal arguments being rehashed in the scientific literature long after the Garner study was published. Apparently, the more outspoken members of

---

* *Caudipteryx*'s hand feathers are symmetrical so they would not have been effective as primary (pinion) feathers extending to the side. The only aerodynamically stable orientation for them would have been straight back, parallel to the air flow. This orientation would make them ineffective as wings but very effective as control flaps, exactly like airplane ailerons.

both sides have such entrenched opinions that they are not very open to conflicting views (a distinctly unscientific attitude).

## Wing-Assisted Incline Running

Ornithologist and flight researcher Kenneth Dial has also tried to bring a fresh idea to the debate. Like James Marden and his stoneflies, Dial observed a behavior in living birds that he decided might be a useful model for the evolution of flapping flight. He noticed that young partridge chicks flap their poorly developed wings when they run up steep slopes. Moreover, they can climb steeper slopes when flapping than when not flapping, and they can climb near-vertical slopes by flapping, even before their wings are big enough for powered flight. Dial dubbed this ability "wing-assisted incline running" or WAIR. He studied WAIR in partridges as they developed from the time of hatching through to fledging, and he also did feather-clipping experiments to adjust wing sizes.[39] He showed that birds could ascend increasingly steeper slopes as their wings got bigger, and he also showed in other experiments that much of the flapping force was directed toward the inclined surface. In other words, WAIR pushes the bird's body against the surface to increase its traction; this requires the flapping wings to produce a downward force component rather than the net upward force they produce during flight.

Dial suggested that flapping flight could have arisen if a small theropod with "incipiently feathered forelimbs"[39] (similar to *Caudipteryx*) gained an advantage avoiding predators or pursuing prey if it could better run up steep slopes. Flapping would thus have evolved at first for improving traction rather than for faster running or for flight. If the protobird also jumped to get down from whatever it ran up, flapping could have also extended its leaps. Once such flapping evolved, converting it to full-fledged flight would just be a matter of enlarging the wings and refining the stroke pattern. Dial and his colleagues have shown how animals can benefit from flapping a small protowing[40] and have analyzed the physics of WAIR in detail.[41,42]

One strong point of WAIR is that even animals with small, poorly feathered protowings can get some advantage from flapping. Also, WAIR starts directly with flapping and bypasses a gliding stage, which some researchers see as a persuasive simplification. Moreover, it sidesteps the traditional controversy because it would work just as well for arboreal or cursorial animals, although it might fit bipedal cursors a bit better. A significant weakness is that WAIR is based on the behavior of animals with shoulders already highly specialized for flapping. Yes, hatchling partridges have small wings with poorly developed feathers, and yes, they can still do something useful by flapping, but their flapping movements use the highly specialized shoulder joint and control system (brain, senses, reflexes) of a fully aerial species. Dial's research group showed that in less aerially skilled birds, like quail, the wingbeat pattern does not change very much between WAIR, level flight, and aerial descent.[43] They suggest that if such a "fundamental stroke" evolved first for WAIR, it could have led to powered flight just by increasing wing size and changing body orientation.

Whether a protobird with an unspecialized shoulder joint and no flapping reflexes could have evolved such a fundamental stroke remains an open question. As with the cursorial flapping-for-thrust scenario, we know of no animals that use WAIR except those that are already flapping flyers, which does not disprove the theory but does show that WAIR is much more difficult to evolve than gliding.

### Directed Aerial Descent

Although originally described in ants, the concept of directed aerial descent that we saw in Chapters 3 and 4 may have a bearing on flight evolution in vertebrates as well.[44] What if arboreal vertebrates evolved skydiving behavior to steer descents or land upright before obvious aerodynamic adaptations appeared? Some arboreal lizards—anoles, for example—use a characteristic spread-eagled posture in falls and make soft, controlled landings even in falls from great heights.[45] (Although he has not published a formal study, a colleague of mine has seen mice do the same thing when falling from a great height.) Given that *Microraptor* was certainly arboreal, and even smaller, feathered, non-avialan theropods like *Eosinopteryx* probably were as well, a trees-down origin for bird flight is no longer incompatible with phylogenies showing birds derived from theropods. Although this idea is not yet mainstream among dinosaur paleontologists, the flight mechanics researchers among my colleagues are all relieved that the more biomechanically reasonable arboreal model can be reconciled with the phylogeny of birds.

Some dinosaur researchers might disagree, but I think that objections to an arboreal origin of bird flight have been laid to rest. The objection that theropods were too big to have climbed trees can no longer be taken seriously. *Compsognathus*, also found in Solnhofen limestone, is only slightly bigger than *Archaeopteryx*, and we now know of several maniraptoran theropods that were even smaller (for example, *Microraptor* and *Eosinopteryx*). The more serious objection that all theropods were bipedal runners and not tree climbers is no longer true, now that we know about *Microraptor* and other similar dinosaurs. I am quite satisfied with the idea of an arboreal evolution of flapping flight occurring in maniraptoran dinosaurs.

## PHYSIOLOGICAL ADAPTATIONS FOR FLIGHT

Birds, like us, are endothermic or "warm-blooded."* Endotherms use a high metabolic rate—essentially a greatly speeded-up physiology—and insulating mechanisms to retain heat and produce an elevated body temperature, even in cold environments. Endothermy comes with a high price; an endotherm typically needs to eat about 10 times as much as an ectothermic ("cold-blooded") animal of the same size. What is the payoff? Physiologists have studied and debated this question for many decades, but most now agree that the main advantage is a great increase in aerobic scope, which means endotherms have much greater endurance, stamina, and continuous muscle power output.[46,47] Living reptiles like lizards are ectothermic and tend to be sprinters. Their muscles can produce short bursts of activity anaerobically (without oxygen), but soon they must stop and endure a period of aerobic recuperation. Birds and mammals, being endotherms, can maintain continuous activity for much longer periods, and their muscles are much better at continuous, long-term power production.

Flight is a very power-hungry activity (Chapter 3), and many biologists think that the increased aerobic scope produced by endothermy may be a requirement for flapping flight in birds. If so, then the question becomes, did protobirds evolve endothermy along with (or soon after) powered flight? Or did certain dinosaurs evolve endothermy first, which then became an exaptation as avialans began to evolve powered flight?

### Endothermic Dinosaurs

When John Ostrom first proposed a close connection between birds and theropod dinosaurs, that linkage led some paleontologists to speculate

---

* Physiologists dislike the term "warm-blooded" because the blood of a basking lizard or dragonfly can be every bit as warm as yours or mine. These ectothermic ("cold-blooded") animals are warmed by the environment, however, not by the high metabolic rate that gives us endotherms an internal furnace.

about the possibility of endothermy in dinosaurs.[48] Prior to that time, people thought of dinosaurs as giant lizards, lumbering and sluggish. The possibility that dinosaurs were endothermic dramatically changed both scientific and public perception, leading to a more dynamic view of dinosaurs being active and birdlike, as exemplified by the "Jurassic Park" book and movies. Although not all scientists are convinced that dinosaurs were endothermic,[49] a consensus has grown that if any dinosaurs were endothermic, theropods are the most likely candidates.

Very large dinosaurs would not have needed endothermy to give them a high and constant body temperature,* but endothermy could have been a great advantage for small theropods such as *Compsognathus* or *Velociraptor*.[50] Small animals lose heat quickly, and if they were not endothermic, these small dinosaurs would not have been able to hunt or run from predators during cold seasons or when temperatures dropped at night. Tiny dinosaurs like *Bambiraptor* and *Eosinopteryx* would have been particularly limited.

Some of the evidence for endothermy in small theropod dinosaurs includes (1) apparent down-like feathers scattered throughout the theropod family tree, even on primitive members such as *Sinosauropteryx*; (2) unequivocal feathers, both downy and pennaceous, on many maniraptorans; and (3) the extremely small size of several recently discovered species, such as *Eosinopteryx* and *Epidendrosaurus*—both are pigeon-sized or smaller. The most obvious function of filamentous or downy feathers is for insulation, and only endothermic animals need insulation. These facts all suggest that endothermy evolved well before flight. The high activity levels, endurance, and strong, continuous muscle power of small theropods would have been a key exaptation in the evolution of bird flight.

### Origin of the Unique Bird Lung

Modern birds have the most sophisticated, efficient respiratory systems of any air-breathing vertebrates. Bird lungs have a unique, one-way, flow-through arrangement. Rather than using in-and-out or tidal air movements into blind-ended pouches or sacs (like us), air flows through bird lungs from back to front in a nearly continuous stream. Again unlike other

---

* Due to their very low surface-to-volume ratio, large dinosaurs would lose heat very slowly and would have had a high and constant body temperature from size alone, with no need for elevated metabolic rates. Scientists call this "gigantothermy" and it would have made *Tyrannosaurus* a fast, active predator even without the physiological modifications for endothermy.

vertebrates, the bird lung does not expand and contract during breathing. Instead, air is driven through the lungs by the bellows-like action of a set of internal air sacs. Bird breathing inflates and deflates these sacs in a coordinated sequence that keeps air flowing smoothly and almost continuously through the lungs. This arrangement is both physically and physiologically more efficient than mammal lungs. It is physically more efficient because much less energy is wasted reversing the direction of air flow, and it is physiologically more efficient because oxygen is easier to extract from a one-way, continuous flow of air than from air that is essentially standing still in the deepest air sacs of mammalian lungs.

And birds do seem better able to obtain oxygen than mammals. In a famous experiment, physiologist Vance Tucker put a mouse and a sparrow in a pressure chamber and lowered the air pressure. When the pressure got so low the mouse was near comatose from lack of oxygen, the sparrow could still fly![51] Scientists thus tend to assume that the bird respiratory system is an adaptation for flying at high altitudes where the oxygen concentration is low, but the real advantage is probably that this unique respiratory system can provide the very high oxygen uptake needed to fuel the high power demands of flapping flight. The high-altitude performance is most likely a beneficial byproduct.

When did the avian respiratory system evolve? Did dinosaurs have one-way lungs and air sacs? Some of the early dinosaur-endothermy proponents suggested that dinosaurs had one-way lungs well before flight evolved, although other scientists see the size of the air passages or the lack of a nasal heat exchanger as evidence against this theory.[52] One thought-provoking study recently showed that living crocodilians have lungs with small regions that seem to have one-way flow.[53] Crocodiles are archosaurs, which makes them both the closest living relatives of birds and fairly close relatives of dinosaurs. One-way lungs in crocodilians raises the surprising possibility that at least partial one-way, flow-through lungs may be primitive in dinosaurs, and the fully developed bird-type system may have evolved in early theropods, long before flight evolved. Indeed, an ectothermic dinosaur would gain little benefit from a highly efficient respiratory system, so if early theropods did have one-way lungs, logic dictates that they would also have been endothermic.

## AVIAN POWERED FLYERS ARRIVE

Living birds have several anatomical modifications to aid flight. I have already mentioned the specialized shoulder joint. This joint has the socket reoriented upward so birds can flap their wings well above the horizontal—many birds can raise their wings so high they nearly or actually touch at the top of the upstroke. (Try lying face-down on a bench and raising your arms to the side; if you can get them much above horizontal, you are doing better than me.) The bird shoulder also includes a unique pulley-like arrangement that allows the tendon of a breast muscle to cross over the top of the shoulder joint and pull up on the humerus (upper arm bone).* This arrangement puts the primary upstroke muscle low on the chest rather than above the shoulder, which keeps the center of gravity of the body in a lower, more stable location. Moreover, this upstroke muscle also twists the wing leading-edge-up, which automatically changes the wing from the downstroke to the upstroke orientation. This motion is especially important in slow flight and during takeoff and landing.

**Pneumatic Bones**

Many people probably have heard that birds have air-filled or pneumatic bones, which are usually described as a flight adaptation to lighten the skeleton. While true, that description is an oversimplification. (For example, we have sinuses—air-filled spaces in our skulls—that have nothing to do with flight.) The air spaces in modern bird bones are tied into the air sac system that supplies the lungs. These air spaces are not respiratory since air spaces in bones can't be used as bellows for pumping or to absorb useful amounts of oxygen. In larger birds, the lightening of the skeleton may be vital for flight, but the benefit is less in smaller birds. Indeed, some small birds as well as diving birds that swim underwater have completely lost skeletal air spaces.[54]

As for the evolution of pneumatic bones, several theropod skeletons have holes that paleontologists see as evidence of air spaces in neck or sometimes back vertebrae; since this feature is more common in larger dinosaurs, it might serve as a weight-reduction adaptation for very large body size.[55] Some researchers see pneumatic bones as evidence that respiratory air sacs and possibly one-way lungs were widespread throughout theropods,[56] although not everyone is convinced that those bones were really air-filled. Even if some or all of those theropod bones really did have air spaces, they do not necessarily require a birdlike respiratory system; they could be connected to other airways, like our sinuses are. Moreover,

---

* The triosseal canal through the shoulder bones serves as the pulley for the tendon from the supracoracoideus muscle.

if pneumatic bones in large dinosaurs are an adaptation for large size, then the ancestors of birds (probably quite small theropods) may have evolved them independently. At this stage, the evidence for when the bird-like, one-way respiratory system evolved is mixed: almost surely by the time fully powered flight evolved and possibly quite early in the evolution of theropods. The evidence of limited one-way flow in crocodilians is intriguing, but that lineage shows no evidence of air sacs or pneumatic bones, so all it really tells us is that dinosaurs may have started out with partial one-way flow in their lungs. A fully developed one-way system must have evolved sometime after the lineage leading to dinosaurs split from that of the crocodilians.

**Other Skeletal Specializations**

The backbone of birds is heavily modified for flight, even compared to non-avialan theropods. It is shortened and stiffened in the trunk region to keep the body's center of gravity close to the wings so the wings can better support the weight of the hindquarters. This shortening is even more extreme for the tail. Whereas non-avialan theropods typically have long, whip-like tails with over 30 vertebrae, living birds have a stubby little stump, irreverently called the "parson's nose." This is supported by a single bone formed from a handful of fused vertebrae, called the pygostyle. When we think of a bird's tail, almost all of what we see as tail is made up of feathers. The flesh and blood part makes up very little of the tail surface. This arrangement both lightens and shortens the tail. Lightness is obviously beneficial for flight, but length is more complex. Long tails can add stability, possibly beneficial for primitive flyers, but living birds have mostly traded stability for maneuverability, so they evolved short tails for enhanced maneuvering.[57]

The hand and wrist skeleton of living birds is heavily modified for flight. Most non-avialan theropod dinosaurs had three or more wrist bones and three (sometimes four) long, clawed fingers of several bones each. Recall from Chapter 2 that living birds have very reduced wrist and hand skeletons, including a single wrist bone, a thumb with only one bone, and one large and one small finger bound tightly together with soft tissue forming a single functional digit (and no claws). Although reduced in number, the bones are relatively stout so they serve as solid anchors for the feathers that bear the body weight in flight (see Fig. 2.1).

Finally, modern birds have a lightweight beak of horn-like material instead of teeth. Beaks appear to be yet another weight-reducing adaptation, presumably for flight. Oddly enough, many extinct birds had teeth whereas beaks apparently evolved independently several times among extinct birds, as well as multiple times among dinosaurs and other reptiles.

### *Archaeopteryx*: A Mosaic

The skeleton of *Archaeopteryx* actually does not show some of the flight adaptations we have discussed. The shoulder socket is only partly reoriented and does not have the pulley arrangement for an upstroke muscle on the chest. *Archaeopteryx* probably could raise its wings only slightly above the horizontal, which suggests it may have had difficulty taking off from level ground or flying slowly. Its backbone is not noticeably shortened and it had a long, bony tail instead of a pygostyle. It apparently had some pneumatic neck vertebrae, but scientists disagree about whether any other bones had air spaces;[55,58] the fossils just have too little detail to say for sure whether it had air sacs and one-way lungs. And, of course, it had teeth. The basal avialans, such as *Archaeopteryx* and *Aurornis*, had pennaceous wing feathers and more or less limited flight abilities, but they also had teeth, long tails, separate fingers with claws, and only partially modified shoulder joints. They were clearly not as specialized for flight as modern birds. So when did these flight-related modifications appear?

### Other Ancient Birds

One of the most primitive avialans to show modern flight adaptations is *Confuciusornis sanctus* (encountered earlier when I described Liaoning fossils with feather traces). *Confuciusornis* is amazing in several ways. It is probably the most abundant species from Liaoning—museums in China apparently house nearly 1,000 specimens, and local farmers have excavated and sold many more to collectors. Many specimens have complete skeletons and soft tissue including well-preserved feathers, and whereas most have very short tail feathers, a few individuals have a pair of long, ribbon-like tail feathers longer than the rest of the body. *Confuciusornis* had very long, asymmetrical primary feathers, giving it unusually high aspect ratio wings and indicating economical flight. It had a horny beak, a pygostyle, and a reoriented shoulder socket, but probably not a triosseal pulley. The skull was heavily built compared to more modern bird skulls, and the breastbone (where the main flight muscles are anchored) was bigger than that of *Archaeopteryx* but not quite as big as those of modern flying birds. It was probably a stronger and more agile flyer than *Archaeopteryx* but probably not as strong or agile as most living birds.

Although *Confuciusornis* would have looked quite birdlike in life, due to its beak and modern-looking feathers, recent phylogenetic studies place it on a primitive side branch near the base of the avialan lineage (Fig. 6.5). After the *Confuciusornis* side branch split off, the two major bird lineages evolved: first the Enantiornithes, or "opposite birds," and then the Ornithurae, the lineage that leads to modern living birds (Neornithes).

Fully powered flight was largely perfected by the time the Enantiornithes split off from the Ornithurae in the very early Cretaceous (over 130 million years ago), although lots of evolution and diversification took place in both lineages after the split. The Enantiornithes seem to have gotten the earlier start. Their fossils have been found all over the world,[59] suggesting they were powerful, effective flyers. Many have perching adaptations, suggesting they were forest dwellers. The Ornithurae were also present in the Cretaceous, but ornithurine fossils are much less common and they seem to have been restricted to oceanic or shoreline habitats. Ironically, the most common Cretaceous ornithurine fossils are of a large, flightless, toothed diving bird called *Hesperornis*, although a few flying species are also known from Cretaceous fossils.[60]

At the end of the Cretaceous, enantiornithine birds disappeared along with the dinosaurs, and the ornithurines survived to give rise to the Neornithes, the modern living birds. The fossil record is not complete enough to show whether the Enantiornithes were gradually being replaced by the Ornithurae before the end of the Cretaceous or whether they disappeared abruptly like the non-avian dinosaurs. In any case, only the ornithurines survived and they have since undergone a stunning diversification. Scientists have described more species of living birds—over 10,000—than all other air-breathing vertebrates—amphibians, non-bird reptiles, and mammals—combined. Birds are clearly the most diverse, widespread, and abundant of the living land vertebrates. Flight, aiding the ability to spread into new environments and exploit new habitats, must surely have contributed to the great ecological success of birds.

**REFERENCES**

1. R. J. Wootton and C. P. Ellington (1991) in *Biomechanics in Evolution*.
2. D. A. Grimaldi and M. S. Engel (2005) *Evolution of the Insects*.
3. J. Evans (1865) *Natural History Review*, series 2.
4. C. Giebel (1877) *Zeitschrift für die Gesammten Naturwissenschaften*.
5. C. Darwin (1860) *On the Origin of Species*.

6. P. Shipman (1998) *Taking Wing: Archaeopteryx and the Evolution of Bird Flight*.
7. J. R. Speakman (1993) *Evolution*.
8. G. Heilmann (1927) *The Origin of Birds*.
9. A. D. Walker (1972) *Nature*.
10. L. D. Martin, J. D. Stewart, and K. N. Whetstone (1980) *Auk*.
11. H. N. Bryant and A. P. Russell (1993) *Journal of Vertebrate Paleontology*.
12. P. J. Makovicky and P. J. Currie (1998) *Journal of Vertebrate Paleontology*.
13. J. A. Gauthier (1986) in *The Origin of Birds and the Evolution of Flight*.
14. U. M. Norberg (1990) *Vertebrate Flight: Mechanics, Physiology, Morphology, Ecology and Evolution*.
15. J. H. Ostrom (1975) *Annual Review of Earth and Planetary Sciences*.
16. J. H. Ostrom (1976) *Biological Journal of the Linnean Society*.
17. W. J. Bock (1965) *Systematic Zoology*.
18. W. J. Bock (1969) *Annals of the New York Academy of Sciences*.
19. A. Feduccia (1996) *The Origin and Evolution of Birds*.
20. U. M. Norberg (1985) in *The Beginnings of Birds*.
21. J. M. V. Rayner (1985a) in *The Beginnings of Birds*.
22. J. M. V. Rayner (1988) *Biological Journal of the Linnean Society*.
23. J. M. V. Rayner (1989) in *Evolution and the Fossil Record*.
24. J. M. V. Rayner (1985b) in *The Beginnings of Birds*.
25. P. Burgers and L. M. Chiappe (1999) *Nature*.
26. Q. Ji, P. J. Currie, M. A. Norell, et al. (1998) *Nature*.
27. P. J. Chen, Z. M. Dong, and S. N. Zhen (1998) *Nature*.
28. T. Lingham-Soliar (2003) *Naturwissenschaften*.
29. A. Feduccia, T. Lingham-Soliar, and J. R. Hinchliffe (2005) *Journal of Morphology*.
30. P. Godefroit, A. Cau, H. Dong-Yu, et al. (2013) *Nature*.
31. D. E. Alexander (2002) *Nature's Flyers: Birds, Insects, and the Biomechanics of Flight*.
32. X. T. Zheng, Z. H. Zhou, X. L. Wang, et al. (2013) *Science*.
33. G. Byrnes and A. J. Spence (2011) *Integrative and Comparative Biology*.
34. M. S. Y. Lee and T. H. Worthy (2012) *Biology Letters*.
35. G. S. Paul (2002) *Dinosaurs of the Air: The Evolution and Loss of Flight in Dinosaurs and Birds*.
36. D. Hu, L. Hou, L. Zhang, et al. (2009) *Nature*.
37. T. Lingham-Soliar (2010) *Journal of Ornithology*.
38. J. P. Garner, G. K. Taylor, and A. L. R. Thomas (1999) *Proceedings of the Royal Society of London Series B*.
39. K. P. Dial (2003) *Science*.
40. K. P. Dial, R. J. Randall, and T. R. Dial (2006) *BioScience*.
41. M. W. Bundle and K. P. Dial (2003) *Journal of Experimental Biology*.
42. B. W. Tobalske and K. P. Dial (2007) *Journal of Experimental Biology*.
43. K. P. Dial, B. E. Jackson, and P. Segre (2008) *Nature*.
44. R. Dudley, G. Byrnes, S. P. Yanoviak, et al. (2007) in *Annual Review of Ecology, Evolution, and Systematics*.
45. J. A. Oliver (1951) *American Naturalist*.
46. A. F. Bennett and J. A. Ruben (1979) *Science*.
47. A. Clarke and H. O. Pörtner (2010) *Biological Reviews*.
48. R. T. Bakker (1972) *Nature*.
49. R. A. Thulborn (1973) *Nature*.

50. F. Seebacher (2003) *Paleobiology*.
51. V. A. Tucker (1968) *Journal of Experimental Biology*.
52. J. A. Ruben, T. D. Jones, and N. R. Geist (2003) *Physiological and Biochemical Zoology*.
53. C. G. Farmer and K. Sanders (2010) *Science*.
54. P. M. O'Connor (2009) *Journal of Experimental Zoology Part a-Ecological Genetics and Physiology*.
55. R. B. J. Benson, R. J. Butler, M. T. Carrano, et al. (2012) *Biological Reviews*.
56. R. J. Butler, P. M. Barrett, and D. J. Gower (2012) *PLoS One*.
57. J. Maynard Smith (1952) *Evolution*.
58. P. Christiansen and N. Bonde (2000) *Proceedings of the Royal Society B-Biological Sciences*.
59. L. M. Chiappe and C. A. Walker (2002) in *Mesozoic Birds: Above the Heads of Dinosaurs*.
60. L. D. Martin (1983) in *Perspectives in Ornithology*.

CHAPTER 7

# Bats

*Wings in the Dark*

People often find songbirds appealing, but most people seem to be repelled by bats. I suspect this reaction to bats is because they are nocturnal, thus creatures of darkness and mystery. Because bats mostly fly in the dark, few people are aware that they are actually a very widespread and successful group. Among mammals, only rodents claim more species than bats, and bats are quite cosmopolitan, inhabiting all continents but Antarctica, from the tropics to the boreal forests. Flight clearly played a role in their dispersal. For example, we know bats reached Hawaii from North America without human help, an inconceivable feat for a tiny non-flying mammal.

Biologists have long recognized a division between the Old World flying foxes and fruit bats ("megabats") and all the other bats ("microbats").* Microbats are what those of us from the temperate regions think of when we think of bats. Microbats tend to be much smaller, they are nocturnal and use sophisticated echolocation, most eat insects, and they are found all over the world. Megabats consist of barely 200 species (compared to approximately 1,000 species of microbats). They are on average larger than microbats, and the group includes all the largest bats. Most megabats don't use echolocation, they are active during the day rather than at night,

---

* Megabats were traditionally placed in the suborder Megachiroptera and microbats in Microchiroptera. As we'll see below, this classification is now obsolete so I have opted not to use the technical terms. I will instead use the informal terms "megabat" and "microbat" as a convenient way to distinguish the Old World fruit bat lineage from the rest of the bats.

and they eat fruit rather than insects. Finally, megabats have a more limited distribution, being restricted to the Old World tropics and subtropics. Nevertheless, megabats and microbats are structurally quite similar; although in the past some researchers have questioned their relationship, they are almost certainly from a single lineage.

Bats unfortunately share parallels with insects when it comes to the fossil record of flight evolution: the oldest bat fossil is from an animal that was already a fully powered flyer. Bats' lightweight skeletons are so delicate that they don't leave nearly as many decent fossils as other mammals, so they have a poor fossil record. Moreover, scientists have nothing like *Archaeopteryx* for bats, so fossils tell us neither anything about how flight evolved, nor about the identity of the direct ancestors of bats.

## CLOSEST RELATIVES?

Biologists would love to be able to say, "The closest relative of bats are the [insert name of group here]." After all, bats were the most recent animals to evolve powered flight, and living mammal groups diversified fairly recently, at least compared to birds and insects. Unfortunately, bats' anatomy is highly specialized. For example, bat wings are highly modified front limbs. As we saw in Chapter 2, each wing has hyper-elongated fingers supporting a membrane formed by a fold of modified skin. Also, bats' ears (both external and internal) and their vocal apparatus are modified for echolocation, an unusual sensing mechanism we will look at later. These specializations have hindered scientists' efforts to pin down the location of bats on the mammal family tree.

Traditional phylogenies based on anatomical characteristics often placed bats in a lineage with colugos* and primates called Archonta (for example, see Fig. 7.1).[1] These phylogenies unite bats with primates based on muscle and reproductive system anatomy and with colugos based on inner ear and forelimb (wing) anatomy.[2] Colugos glide on a wing membrane made of a sheet of skin stretched between front and hindlimbs. Unlike all other gliding mammals, colugos have webbed fingers, so the hand acts as part of the wing (see Chapter 4, Fig. 4.5). Given that much of the bat wing is supported by modified fingers, colugo wing anatomy thus makes a nice model for the early evolution of bats from gliding ancestors.[3] If bats are

---

* Colugos are the large, Southeast Asian gliding mammals sometimes called "flying lemurs" that we encountered in Chapter 4.

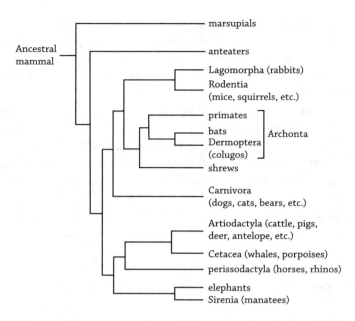

**Figure 7.1:**
Traditional mammal phylogeny based on anatomical features (simplified from Novacek;[1] some minor groups omitted for clarity).

closely related to colugos, then the evolutionary sequence from a colugo-like glider to powered flight seems logical and self-evident.

Unfortunately, more recent phylogenies based on genes don't support this neat picture. These molecular phylogenies keep primates and colugos on the same lineage, but bats fall far from this reduced archontan group (Fig. 7.2). As of this writing, the position of bats on various molecular phylogenies has not completely stabilized. These trees generally show bats in or close to the Laurasiatheria, a lineage including whales, cloven-hoofed animals (cattle, deer, etc.), horses, and carnivores (dogs, cats, bears, etc.). Whether bats are nested within Laurasiatheria, or are at the base of the lineage, or just outside the group remains to be seen, and no consensus yet exists on which group represents the closest relatives of bats.[4-6] Some phylogenies even show shrews as closely related to bats.[7,8] This possible relationship is intriguing because shrews also use echolocation, although their system is apparently not nearly as sophisticated as that of bats (see Box 7.1. Echolocation in Context: The Sensory Suite of Bats).[9,10]

The fossil record is not much help in teasing out bat evolutionary relationships. Scientists have described at least two very complete bat fossil skeletons from the early Eocene (50 to 55 million years ago). *Icaronycteris index* was discovered in the 1960s and studied by different researchers

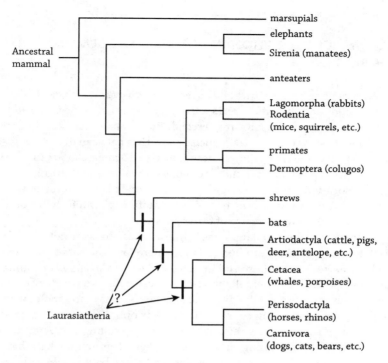

**Figure 7.2:**
A molecular (gene-based) phylogeny of the major mammal groups. Various studies differ in whether shrews and bats should be included in Laurasiatheria. This is a somewhat simplified composite version of trees in Novacek[4] and Springer et al.[5]

over the years.[11,12] This species had fully developed wings, spanning 37 centimeters (14½ inches), that were anatomically indistinguishable from those of living bats; based on the parts of the skull housing the inner ear, it also used sophisticated, insect-detecting echolocation. Researchers recently discovered the fossil skeleton of *Onychonycteris finneyi* in the same fossil beds and from approximately the same age as *Icaronycteris*.[13] Although these two species lived at about the same time, *Onychonycteris* is clearly more primitive: it had proportionally shorter wings than all other bats, and it had claws on all of its fingers.* Moreover, its inner ear does not show the specializations for echolocation typical of most bats. Nevertheless, *Onychonycteris* was clearly a bat and was very specialized for flight compared to non-flying mammals. Evidence from the fossils as well as

---

* Living bats only have claws on the thumb, or in a few cases, on the thumb and index finger.

> *Box 7.1:* ECHOLOCATION IN CONTEXT: THE SENSORY SUITE OF BATS
>
> Even though microbats are highly specialized to use sensitive, sophisticated echolocation, bat sonar is a distinctly short-range sense and is fatiguing to use while bats are perched. Both behavioral experiments and the physics of calling frequencies show that bats can detect insect-sized objects only when they are less than 5 or 10 meters (15 to 30 feet) away. Bats typically fly around 3 to 5 meters per second (7 to 11 mph), so echolocation detects prey only 1 or 2 seconds before interception. Although bats always seem to use echolocation to catch flying prey, they also make extensive use of other senses.
>
> Contrary to the popular phrase "blind as a bat," bats are not blind. In fact, when researchers studied vision in microbats, the bats turned out to have better vision than similar-sized rodents. (Megabats, of course, are diurnal and have excellent vision.) Researchers have discovered that when there is enough light to see obstacles, bats stop echolocating and may even use vision to search for insects on foliage or branches. If background noise is quiet enough, they may stop calling and listen passively for insects rustling in leaf litter. Several groups of bats have evolved to eat pollen and drink nectar, and these bats rely heavily on their sense of smell to find food.
>
> Bats apparently evolved their advanced sonar to catch insects in flight, perhaps based on an earlier, simpler system to avoid obstacles while flying in the dark. Bats can use echolocation for sensing their surroundings, but they are not exclusively dependent on sonar; they have other well-developed senses as well. This sensory flexibility undoubtedly contributed to bats' success and diversity.

some molecular phylogenies suggest that bats split off from other mammals well before the Eocene, back into the Cretaceous before the dinosaur extinction. The great antiquity of this split is one of the reasons that scientists have had difficulty discovering bats' nearest relative. The ancient split means that their closest relatives among living animals probably diverged quite a bit during this long period and may no longer share many traits with bats.

Most of a bat's skeleton is heavily modified for flight, and these modifications make bat fossils instantly recognizable. If we had the fossil skeleton of a bat ancestor from before flight evolved, would we recognize it as being related to bats? I posed this question to a colleague who studies bat biology. He thought such an ancestor would probably be recognizable from skull anatomy, but without the skull, making a connection to bats might be quite difficult. Lacking any sort of transition fossil between *Onychonycteris* and such a possible pre-flight ancestor, and also lacking the ability to extract genes and construct molecular phylogenies for such fossils, scientists would be hard-pressed to recognize the evolutionary relationship of a non-flying ancestor with bats.

## FLIGHT EVOLUTION

Both the molecular phylogenies and the fossils suggest that bats evolved flight near the time of the end-Cretaceous extinction event (65 million years ago), meaning that bats evolved flight long after birds had evolved powered flight and diversified and quite likely while pterosaurs (Chapter 8) were still around. The living lineages of bats seem to have arisen near or just before the end of the Cretaceous,[14] which implies that the entire bat lineage and flight must have arisen even earlier. Whether flying bats had just recently evolved or been around for a long time at the end of the Cretaceous, the evidence is too sparse to say.

### Unquestionably Arboreal

In stark contrast to the situation with bird research, scientists are in near-universal agreement that bats evolved from arboreal, gliding ancestors,[15-19] as first proposed by Darwin.[20] The reason is simple: bat legs help support the wing membrane, so how could a running protobat's legs become so thoroughly incorporated into the wing? Fast running is a critical component of the cursorial model, and the basic structure of the bat wing (not to mention the highly modified legs and backward-facing feet) seems to preclude running. Most bats are reasonably good climbers; many roost in trees or tree-holes, and out of all the thousand or so known species of bats, only a handful voluntarily spend much time walking on the ground. Moreover, for decades, scientists considered colugos to be the closest relatives of bats,[1,21,22] which suggested that their common ancestor was a glider. Nowadays, as genetic studies provide increasingly robust evidence that bats and colugos are not closely related, we think these two animal groups must have evolved their wings independently. Nevertheless, colugos still make a good model for what ancestral protobats must have looked like. Like bats, colugos have a wing membrane formed from layers of skin supported by front and hindlimbs, are awkward on the ground, and are largely nocturnal. Indeed, without ever mentioning colugos explicitly, biologist James D. Smith proposed an evolutionary sequence from an arboreal leaper to bats in which the wing he showed for the intermediate glider (Fig. 7.3) looks identical to a colugo wing.[3]

**Yet Some Raise Questions**

A few researchers interested in the evolution of flight in both birds and bats have raised questions about the gliding-to-flapping transition, as we saw in Chapter 3 and Chapter 6. Some assert that wings specialized for gliding would not produce useful force when weakly flapped, and even that flapping such a wing would cause it to perform more poorly than during gliding. I am puzzled why this claim persists in the scientific literature, given that it has repeatedly been refuted both theoretically and empirically.[23-26] Others have argued that flapping a wing designed for gliding would lead to instability and loss of control. This remains to be tested and will most likely require something like model tests in a wind

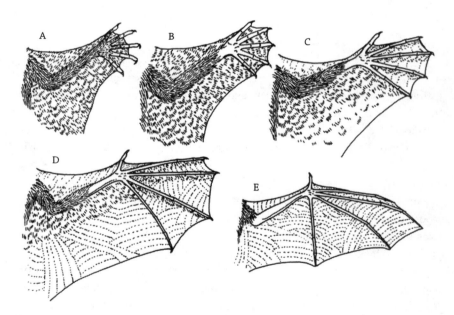

**Figure 7.3:**
The evolutionary sequence proposed by James Smith from an arboreal glider to a bat; letters indicate order of the evolutionary sequence.[3] The early stages, particularly stage B, look very similar to a colugo wing. (Used with kind permission of Springer Science + Business Media.)

tunnel to confirm or refute. In any case, no one so far has proposed a reasonable competing theory for flight origins in bats, so scientists generally accept the arboreal model as the most likely route for achieving powered flight in bats.

## HOW MANY FLIGHT ORIGINS?

Biologists have traditionally separated megabats from microbats within the bat order, Chiroptera. Megabats are limited to the Old World tropics and consist of a single family, Pteropodidae, versus about 16 families of microbats spread around the world. Megabats are mostly diurnal fruit eaters whereas microbats are largely nocturnal insect eaters. While the groups are clearly different, the terms "megabats" and "microbats" are somewhat misleading. In fact, as shown in Figure 7.4, the size ranges of megabats and microbats overlap substantially, although megabats on average are significantly larger than microbats. These two groups were originally considered to be separate suborders—different lineages—of bats, but we now know that bat relationships are a bit more complex (as we'll see).

### One Origin or Two for Bats?

Despite their differences, megabats and microbats are so similar in body form that for over a century, scientists took for granted that they were both part of one lineage ("monophyletic"). Few scientists took seriously suggestions that megabats and microbats might have evolved independently.[27] That all changed in the 1980s when John D. Pettigrew and colleagues published studies showing that the anatomy of the part of the brain that processes vision (and the way the eyes are connected to that brain region) is very similar in megabats, colugos, and primates but different in all other mammals, including microbats (see Box 7.2. Megabat, Colugo, and Primate Vision: Evolutionary Red Herring).[28,29] They interpreted their results to mean that megabats are close relatives of primates but microbats are not. That in turn would mean that powered flight must have arisen twice, once in megabats and once in microbats. They went on to develop a phylogeny based on nervous system characteristics that placed megabats close to cologos and primates and positioned microbats as the lineage that split off earliest among the dozen or so mammal groups they included.[30] Their argument is bolstered by other

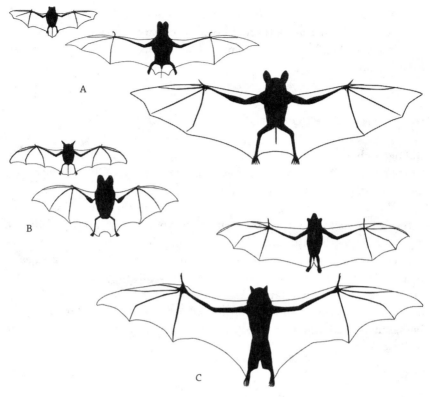

**Figure 7.4:**
Silhouettes of bats with extended wings, drawn to scale, to show the relative sizes of some microbats and some megabats. The smallest bat shown here, *Pipistrellus pipistrellus* (upper left), has a wingspan of about 23 centimeters (9 inches); although on average, microbats are smaller than megabats, this diagram shows how much their size ranges overlap. A. Microbats of the Yangochiroptera lineage. B. Microbats of the Yinpterochiroptera lineage. C. Megabats (Yinpterochiroptera). (Courtesy of S. T.)

differences: for example, all microbats use echolocation and produce the sound pulses using their larynx (voice) whereas very few megabats use echolocation and those that do produce sound pulses by clicking their tongues.

The anatomical evidence is fairly ambiguous, and other researchers pointed to strong similarities in forelimb (wing) structure shared by all bats.[31] Proponents of separate origins countered that this similarity is to be expected if the wings of megabats and microbats were subjected to similar, flight-related selection pressures and physical constraints. Single-origin proponents countered in turn that even though some of the fine anatomical details of wing structure differ, the underlying nerve

> *Box 7.2:* MEGABAT, COLUGO, AND PRIMATE VISION: EVOLUTIONARY RED HERRING
>
> John Pettigrew did not base his phylogeny uniting megabats (but not microbats) with colugos and primates on incorrect characteristics: megabats, colugos, and primates share a set of visual and vision-related brain features that are very unusual among mammals. These animals possess color vision, common among birds but rare among mammals. They also have eyes facing forward with overlapping right and left visual fields, which gives them binocular vision. Animals use binocular vision for depth perception—judging distances—but these three groups have taken it a step further. They have modified the neural wiring that connects the eyes to the brain in such a way that both eyes send part of their signals to both sides of the brain. This improves the brain's ability to compare right and left images and enhances depth perception. If megabats, colugos, and primates did not inherit these features from a common ancestor, how did they come to be so similar? Science cannot give a definitive answer, but we do have some clues.
>
> All three of these groups apparently had a very early history of being diurnal (day-active), tropical, arboreal fruit eaters. Color vision is easy: color vision lets animals tell from a distance if many kinds of fruit are ripe (color being a signal plants probably evolved initially to help fruit-eating, seed-dispersing birds). Depth perception would also be at a premium for animals that make their living leaping—or gliding or even flapping—from one small branch to another high in the tops of very tall rain forest trees. Reaching out to grab the landing target but missing it could have terminal consequences. Perhaps depth perception was also beneficial when reaching out to pluck a ripe fruit (in the tops of those very tall trees). Our best evidence suggests that megabats and the colugo-primate lineage independently evolved rewiring of the visual pathways as a common response to improve depth perception starting from the standard mammalian pattern. Similar ecological and physical constraints often lead to convergent anatomies, such as body forms of sharks and porpoises, although convergence of neural pathways seems to be rare (but is not well-studied).

innervation pattern for the flight muscles is the same in both groups.[32] This controversy grew rather heated: either the visual systems of megabats and primates evolved convergently or the wings of megabats and microbats evolved convergently, but anatomical evidence was inadequate to resolve the argument.[17,31,33–35]

### Genes Say "Single Origin"

As researchers began to apply molecular phylogenetic methods to bats and began building phylogenies based on genes, a consensus began to emerge from the chaos. Some of the earliest studies firmly supported a single bat lineage,[36] which Pettigrew and supporters of two origins for bats criticized on rather technical chemical and statistical grounds. The tide shifted decisively when Irish researcher Emma Teeling and colleagues published a series of studies looking at many genes in many species of bats, as well as in primates (humans), colugos, dogs, and mice.[6,37,38] These studies clearly showed that megabats and microbats are each other's closest relatives, and neither has any close relationship to primates or colugos. These "gene trees" thus leave no doubt that powered flight evolved only once in bats, and while the ancestor of bats may have looked somewhat like a colugo, bats did not arise from the same lineage as colugos (Fig. 7.5).

The molecular studies also showed something quite unexpected: microbats do not form a single lineage because some microbats are actually more closely related to megabats than to other microbats. Ironically, one of the first molecular studies to show this split was one co-authored by John Pettigrew, the "two origin" supporter.[39] Even before the molecular phylogenies showed a deep split in the microbats, at least one bat researcher proposed

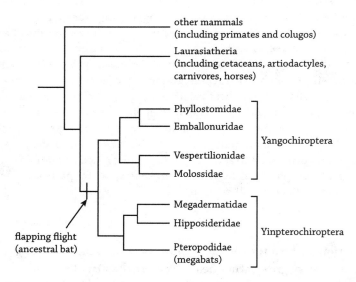

**Figure 7.5:**
This molecular phylogeny of some major bat families shows no close relationship between bats and colugos and implies a single origin for flight in bats. All bat families other than Pteropodidae are microbats. (Simplified from the phylogenies in Teeling et al.)[38]

that microbats should be divided into two distinct lineages. Karl Koopman called these two lineages Yinochiroptera and Yangochiroptera, basing the names not completely on whimsy: yin is the active principle and yang the passive principle in Confucian philosophy, and his Yinochiroptera have a unique mobile bone in their upper jaw.[40] At first, some bat researchers merely saw the molecular phylogenies as confirming Koopman's microbat split. The gene-based trees, however, also generally showed that many microbats (mostly those in Koopman's Yinochiroptera) belonged on the same lineage as megabats. Megabats have long been placed in a single family, the Pteropodidae, so biologists have now combined megabats and their close microbat relatives into a lineage called Yinpterochiroptera.*

The molecular phylogenies answered some questions but raised others. Scientists now agree that powered flight arose only once in bats and that bats are not closely related to colugos (darn, that made such a nice story). But the microbats in Yinpterochiroptera, such as Old World leaf-nose bats (Hipposideridae, Fig. 7.5), are all active and sophisticated nocturnal echolocators like those in Yangochiroptera, whereas megabats are diurnal and very few use echolocation. That means that either (1) bats evolved echolocation very early and megabats lost it; or (2) echolocating yinpterochiropterans evolved echolocation independently from Yangochiroptera. Given how central echolocation is to the life of a nocturnal bat and given its close association with flight in living bats, let's now take a closer look at echolocation in bats.

## HEARING THINGS

Echolocation or sonar† means producing a sound that can be reflected back from solid objects and listening for the echoes to sense where those objects are located. Toothed whales (such as dolphins and porpoises) and microbats have well-developed echolocation systems, and two different kinds of cave-dwelling birds and at least a few species of shrews seem to use it as well.[9,41] Bats and whales have clearly evolved echolocation for

---

* The observant reader might note some redundancy in this new term. "Yinpterochiroptera" translates approximately as "active wing hand wing," and even though some researchers would prefer a less ad hoc term, this name has become widely accepted in the scientific literature.

† "Sonar" usually refers to a human-made device such as used on ships to detect submerged submarines and echolocation usually refers to an animal's sensing mechanism, but the underlying process is identical; some researchers use them interchangeably.

sensing objects when visibility is poor—at night for bats, in murky or very deep water for whales.

Microbats use their voices to produce ultrasonic sounds, that is, sounds with such high frequency (pitch) that humans cannot hear them. The reason bats use such high-frequency sounds is that the wavelength of the sound determines the smallest object that will return an echo, and wavelength is inversely related to frequency: you need a very high frequency to get a short enough wavelength to detect small objects, say, the size of insects. A handy side effect of using such high frequencies is that the sound can be more narrowly focused and directional; in contrast, lower frequency sounds tend to spread out all over. This directionality has obvious advantages for sensing the direction of objects. The high frequencies also have a drawback, however: they attenuate (fade out) very rapidly over distance so bats must make them very loud, and this requires a lot of power. These clicks or sound pulses are so intense that if we could hear them, a bat echolocating at arm's length would be painfully loud.

**Ultrasonic Call Types**

Bats use echolocation mainly for two things: detecting and avoiding objects when they fly in the dark, and insect "hawking" where a flying bat detects and captures a flying insect. One of the earliest studies of bat sensing showed that intact flying bats could avoid obstacles in total darkness but not if their ears were plugged.[42] This ability pales in comparison to bats' ability to detect, track, intercept, and capture insects in flight. Bats have evolved remarkably sophisticated techniques for zeroing in on flying insects. Some use frequency-modulated (FM) or broadband calls that give very precise distance detection whereas others use constant-frequency (CF) calls that carry farther. Constant-frequency calls also allow the bats to use the Doppler shift* of the echo to sense their closing speed on the target insect. Many CF bats can even sense the wingbeat frequency and orientation of an insect and use that information to decide whether to bother chasing it. Moreover, as hawking bats close on a target insect, their calls come faster and faster, so the bat gets more frequent updates of the target's position. Bats that hawk in the open tend to use CF calls whereas bats that hawk in cluttered environments (within a forest) are more likely

---

* The Doppler shift is the familiar change in pitch (frequency) due to motion, such as when a train whistle or siren rises in pitch as it approaches the listener and then falls in pitch after passing and moving away.

to use FM calls. In general, FM bats are more common but some bats use a combination of both call types depending on the situation.

Separating some microbats into the Yinpterochiroptera and others into the Yangochiroptera means that CF calls, as well as other specialized features of echolocation, evolved independently in the two lineages. This in turn means that either (1) basic echolocation is ancestral in all bats, and megabats lost echolocation; or (2) the megabat lineage never had echolocation, and yangochiropterans and small yinpterochiropterans each evolved echolocation entirely independently. In the second case, FM and CF styles are entirely convergent and not inherited from a common ancestor. So far we don't have the evidence to answer this conclusively.

### Echolocation and Flight

What does all this have to do with flight evolution? Flight and echolocation in bats are linked by nocturnality and by a literal physical linkage between the wingbeat and call production.

Bats are probably nocturnal because birds were already well established and diverse when bats were first evolving flight in the mid- to late Cretaceous. The ancestors of bats were most likely already nocturnal and were able to evolve powered flight at night when birds are largely inactive. Some scientists have speculated that being nocturnal allowed bats to avoid competition with birds—although they would not have been competing for food because insect-hawking birds did not evolve until much later. Other researchers have pointed out that the ancestral owls, hawks, and falcons were around when the bats first took to the air, so being nocturnal was probably important to avoid predators.

If you were a small, nocturnal insect eater, having sensitive hearing might have helped you detect and locate the movement noises of prey. This passive acoustic detection could have led to some form of echolocation, which would have been handy for getting around on really dark nights or detecting jumping or flying insects. We don't know whether the non-flying or the gliding ancestor of bats already used some form of echolocation or whether their descendants evolved it later; either way, it would have been a huge benefit to a nocturnal flyer. Many dark-adapted mammals rely on the sense of touch, particularly long whiskers, to find their way in total darkness. For a flyer moving rapidly through the air, whiskers would not have been much use for sensing obstacles in time to avoid them.

In order to make their echolocation calls loud enough to detect insects at a useful distance, bats must put quite a lot of muscle power into echolocating. In fact, when bats use echolocation while perching, their metabolic energy use increases tenfold over resting, meaning that echolocation alone requires muscle power akin to vigorous exercise. Bats' metabolic rates go up even more, by 15 to 20 times the resting rate, when they are flying, but John Speakman and Paul Racey found in their landmark study that echolocation while flying does not further increase a bat's metabolic rate.[43] In other words, while flying, bats get echolocation "for free." When they are searching, they call once per wing beat, so they produce their calls as a byproduct of flight muscle movements.

This tight linkage between echolocation and flight has led at least some researchers to suggest that bats evolved echolocation and powered flight simultaneously.[13] Traditionally, researchers have argued that either echolocation evolved before flight[44] or flight evolved before echolocation.[45] To me, a third possibility seems plausible: the great power requirements of sophisticated, insect-detecting sonar means that as flapping evolved, high-power echolocation became easier and evolved more or less simultaneously. I can even envision that increasing the strength of echolocation calls in a gliding bat ancestor might have stimulated and selected for the earliest, weak flapping.

### Echolocation and Size

One curious feature of flight-powered bat echolocation is that it might limit bat body size. As flapping flyers get bigger, their wingbeat frequency decreases. Scientists have suggested that as the ancestors of megabats increased in size they reached a point where their echolocation calls were not frequent enough to usefully detect prey. In other words, they traveled so far between calls that they could not effectively track prey or detect obstacles in time to avoid them. Large size might also have reduced predation pressure and allowed early megabats to spend more time being active during daylight; in this case, echolocation would not have been as important and so it was eventually lost. By the same token, the flight-echolocation linkage may be the reason the specialized echolocating bats have all remained "micro." If echolocation remains beneficial, these bats may be restricted to small body sizes.*

---

* Speed and maneuverability may affect size as well. Big flyers are faster, which seems like an advantage. This speed difference, however, means that smaller flyers can turn more sharply than big ones; even small bats, as maneuverable as they are, cannot follow all the twists and turns of an evading insect. They must either try to intercept, rather than follow, evasive prey, or choose prey that are less evasive. As they get bigger, large bats may not even be maneuverable enough for interceptions or following less evasive prey.

## FLIGHT SPECIALIZATIONS

Bats evolved major anatomical and physiological adaptations to achieve effective powered flight. The modifications that converted an arm skeleton into a wing skeleton are pretty obvious (recall Fig. 2.2 from Chapter 2). The skin of the wing membrane is also heavily modified and quite different from the skin on the rest of the bat's body (which is typical mammalian skin). The wing skin is much more compliant, meaning it stretches and deforms more easily.[46] Studies by Arnold Song and colleagues on wind-tunnel models with compliant surfaces show that such flexible wings can have higher maximum lift coefficients and can be more stall-resistant than rigid wings.[47] Bat wings also contain a complex array of muscle layers, which we assume are used to adjust the wing's shape throughout the flight stroke, as well as a variety of stretchy or tough reinforcing fibers.[46] In short, bat wings are a lot more complex and sophisticated than meets the eye.

Flight imposes high power requirements and, as we saw for birds with their flow-through lungs, bats have also evolved lung and circulatory system modifications to improve oxygen delivery to active flight muscles. Bats have substantially larger lungs and hearts for their size than other mammals.[48] For example, bat hearts can be up to three times bigger than hearts of other similarly sized mammals. Bats also tend to have more red blood cells and so more oxygen-carrying hemoglobin in their blood. Finally, measuring the metabolic rate of animals in flight is quite challenging, but we have a least a little evidence that bats have higher metabolic scope than other mammals, meaning that they can sustain an unusually high metabolic rate for their size.[49] Unlike birds, with their highly modified respiratory system, bats have just evolved incremental improvements on the typical mammalian system. These changes, however, are clearly sufficient to give bats the necessary power and stamina to be successful powered flyers.

## REFERENCES

1. M. J. Novacek and A. R. Wyss (1986) *Cladistics*.
2. J. Shoshani and M. C. McKenna (1998) *Molecular Phylogenetics and Evolution*.
3. J. D. Smith (1977) in *Major Patterns in Vertebrate Evolution*.
4. M. J. Novacek (2001) *Current Biology*.
5. M. S. Springer, M. J. Stanhope, O. Madsen, et al. (2004) *Trends in Ecology & Evolution*.
6. E. C. Teeling, O. Madsen, R. A. Van Den Bussche, et al. (2002) *Proceedings of the National Academy of Sciences of the United States of America*.
7. R. M. D. Beck, O. R. P. Bininda-Emonds, M. Cardillo, et al. (2006) *BMC Evolutionary Biology*.

8. W. J. Murphy, E. Eizirik, W. E. Johnson, et al. (2001) *Nature*.
9. K. A. Forsman and M. G. Malmquist (1988) *Journal of Zoology* (London).
10. E. Gould, N. C. Negus, and A. Novick (1964) *Journal of Experimental Zoology*.
11. G. L. Jepsen (1966) *Science*.
12. M. J. Novacek (1985) *Nature*.
13. N. B. Simmons, K. L. Seymour, J. Habersetzer, et al. (2008) *Nature*.
14. E. C. Teeling (2009) *Trends in Ecology & Evolution*.
15. U. M. Norberg (1985) *American Naturalist*.
16. J. M. V. Rayner (1989) in *Evolution and the Fossil Record*.
17. N. B. Simmons (1995) *Symposium of the Zoological Society of London*.
18. K. L. Bishop (2008) *Quarterly Review of Biology*.
19. N. P. Giannini (2012) in *Evolutionary History of Bats: Fossils, Molecules and Morphology*.
20. C. Darwin (1860) *On the Origin of Species*.
21. N. B. Simmons and T. H. Quinn (1994) *Journal of Mammalian Evolution*.
22. T. A. Vaughan, J. M. Ryan, and N. J. Czaplewski (2000) *Mammalogy*.
23. U. M. Norberg (1985) in *The Beginnings of Birds*.
24. R. L. Nudds and G. J. Dyke (2009) *Evolution*.
25. K. Peterson, P. Birkmeyer, R. Dudley, et al. (2011) *Bioinspiration & Biomimetics*.
26. N. Vandenberghe, J. Zhang, and S. Childress (2004) *Journal of Fluid Mechanics*.
27. K. J. Jones and H. H. Genoways (1970) in *About Bats: A Chiropteran Symposium*.
28. J. D. Pettigrew (1986) *Science*.
29. J. D. Pettigrew (1991) "Wings or brain . . ." *Systematic Zoology*.
30. J. D. Pettigrew, B. G. M. Jamieson, S. K. Robson, et al. (1989) *Philosophical Transactions of the Royal Society of London B Biological Sciences*.
31. R. J. Baker, M. J. Novacek, and N. B. Simmons (1991) *Systematic Zoology*.
32. J. G. M. Thewisen and S. K. Babcock (1991) *Science*.
33. J. D. Pettigrew (1991) "A fruitful, wrong hypothesis . . ." *Systematic Zoology*.
34. J. D. Pettigrew (1995) in *Ecology, Evolution and Behaviour of Bats*.
35. N. B. Simmons, M. J. Novacek, and R. J. Baker (1991) *Systematic Zoology*.
36. W. Bailey, J. Slightom, and M. Goodman (1992) *Science*.
37. M. S. Springer, E. C. Teeling, O. Madsen, et al. (2001) *Proceedings of the National Academy of Sciences of the United States of Americ*a.
38. E. C. Teeling, M. Scally, D. J. Kao, et al. (2000) *Nature*.
39. J. M. Hutcheon, J. A. W. Kirsch, and J. D. Pettigrew (1998) *Philosophical Transactions of the Royal Society of London B Biological Sciences*.
40. K. F. Koopman (1984) *Bat Research News*.
41. H. A. Thomassen, S. Gea, S. Maas, et al. (2007) *Hearing Research*.
42. J. D. Altringham (2011) *Bats: From Evolution to Conservation*.
43. J. R. Speakman and P. A. Racey (1991) *Nature*.
44. M. B. Fenton, D. Audet, M. K. Obrist, et al. (1995) *Paleobiology*.
45. N. B. Simmons and J. H. Geisler (1998) *Bulletin of the American Museum of Natural History*.
46. S. M. Swartz, M. S. Groves, H. D. Kim, et al. (1996) *Journal of Zoology*.
47. A. Song, X. D. Tian, E. Israeli, et al. (2008) *AIAA Journal*.
48. J. N. Maina (2000) *Journal of Experimental Biology*.
49. S. P. Thomas (1987) in *Recent Advances in the Study of Bats*.

CHAPTER 8

# Pterosaurs

*Bygone Dragons*

Pterosaurs are the last remaining group of powered flyers. I have saved them for last partly because we know the least about them and partly because no one has developed any theories of flight evolution for pterosaurs aside from those we have already encountered.

Pterosaurs have long been objects of rapt public attention, probably because some of them were huge. Indeed, they include the largest flying creatures ever known, although many were no bigger than seagulls. Pterosaurs were reptilian and lived in the Triassic, Jurassic, and Cretaceous periods; many were huge, and they died out with the non-avian dinosaurs. For this reason, many people probably assume pterosaurs were a kind of dinosaur. In fact, while both pterosaurs and dinosaurs (along with birds and crocodilians) are archosaurs, they represent two separate and distinct lineages of archosaurs—in other words, pterosaurs are not dinosaurs. And while the traditional view is that pterosaurs and dinosaurs are each other's closest relatives, not all researchers agree.

Pterosaurs, or "pterodactyls" as they are sometimes known (see Box 8.1), were the first vertebrates to evolve powered flight, following the insects into the air more than 220 million years ago. They are also the only major lineage of powered flyers to have gone entirely extinct, disappearing along with the non-avian dinosaurs at the end of the Cretaceous, about 65 million years ago. Because we have no living pterosaurs to study, we must make many inferences and assumptions about them based on living flying animals that we can study. All modern animals flap to produce thrust and use variations on the same basic stroke for cruising flight, so we are confident that pterosaurs did the same. Unfortunately, pterosaur wings were so

> **Box 8.1:** WHAT ARE THEY CALLED?
>
> Scientists call the extinct flying reptiles that are the subject of this chapter "pterosaurs" and they tend to wince at the common use of "pterodactyl" as a synonym for "pterosaur." What's wrong with using "pterodactyl" as a common name?
>
> The origin of "pterodactyl" goes back to the earliest scientific description of a pterosaur. The original pterosaur fossil was first described as a sea creature, but when Georges Cuvier realized it was actually a flyer, he re-described it and in 1809 gave it the genus name "Ptéro-Dactyle." This name means "wing finger" and refers to the hyperelongated fourth finger that supports most of the wing. The rules for taxonomic naming were just then starting to gain wide acceptance, and under those rules, "Ptéro-Dactyle" became the Latinized genus name "*Pterodactylus*" for that first pterosaur specimen, which we now know as *Pterodactylus antiquus*. Obviously the English common name "pterodactyl" comes directly from Cuvier's original genus name, so it has a long history of common usage. The very fact that Cuvier used it as a technical term, however, causes the problem.
>
> Technically, *Pterodactylus* is a genus, that is, a group of a few closely related species out of the hundred-plus known pterosaur species. Moreover, *Pterodactylus* is the type genus—a sort of exemplar group—for the suborder Pterodactyloidea, a lineage of later, anatomically distinct pterosaurs. To a biologist or paleontologist, "pterodactyl" only refers to the handful of species in the genus *Pterodactylus* and it can also be confused with "pterodactyloid," which refers only to members of one out of several main lineages within the pterosaurs. Thus, to a scientist, "pterodactyl" is ambiguous: is the speaker referring to just the genus, or the suborder, or to all pterosaurs? I can't hope to extinguish the use of "pterodactyl" as a common name, but I do hope readers now understand why scientists much prefer to use "pterosaur" for the name of the entire group.

different from bird or bat wings (see Chapter 2) that scientists don't entirely agree on some of the details, such as how they took off and landed, how efficient their wings were, or what they did with their wings when not flying.

Hard facts about flight evolution in pterosaurs are even skimpier for them than for birds or bats. With living animals, we can use molecular (gene-based) phylogenies to clarify evolutionary relationships, and the anatomy and physiology of living species can give us pretty strong clues about how similar structures worked in their extinct ancestors and relatives. In the case of pterosaurs, no human has ever seen an animal fly by flapping a membranous wing supported by one enormous finger. Similarly, with no live animals to study and almost no information on their soft tissues, anything researchers say about pterosaur physiology is largely informed speculation. And finally, we have no early transition fossils for pterosaurs, and researchers don't entirely agree on what group may have been their immediate ancestors. Nevertheless, they were the first vertebrates to fly under power and they include the largest animals that have ever flown, so they are worthy of our attention.

## ORIGINAL PTEROSAUR FOSSIL DISCOVERIES

Everyone knows what a bird or bee is just from common, everyday experience. And though most people may never actually see a bat, they will still know of bats from common cultural references ("blind as a bat"). So when researchers find a fossil bird or bat, they have a reasonably good idea what kind of animal they are dealing with. Pterosaurs, in contrast, are not part of our common, everyday experience. When scientists first encountered pterosaur fossils over 200 years ago, they came up with a variety of quite different interpretations for them.*

The first scientifically described pterosaur fossils were from the same Solnhofen limestone that produced the *Archaeopteryx* fossils, but the first pterosaur was found over a half century before *Archaeopteryx*, in the late 1700s. Cosimo Collini, the first scientist to formally describe a pterosaur fossil (Fig. 8.1), thought it was some sort of strange sea creature.[1] The great French anatomist Georges Cuvier, however, immediately realized it was a flying animal;[2] he later placed it in the genus *Ptéro-Dactyle*,[3] which became the basis for the modernized species name, *Pterodactylus antiquus* (see Box 8.1). Although Cuvier was well aware that this pterosaur was a reptile, Samuel Thomas von Soemmerring misidentified a specimen of *Pterodactylus* as a strange bat that he thought represented a transition between birds and bats. He published a picture of the animal reconstructed in a very bat-like posture,[4] as shown in Figure 8.2. Although Cuvier and others soon pointed out Soemmerring's error and firmly established that pterosaurs were reptilian and not bats, Soemmerring's image lingered in the popular imagination. Indeed, Kevin Padian claims that Soemmerring's bat-like

---

* The history of the earliest studies and diverse interpretations of the first pterosaur fossils is recounted in fascinating detail in Peter Wellnhofer's *Illustrated Encyclopedia of Pterosaurs*, later reprinted in the United States as *The Illustrated Encyclopedia of Prehistoric Flying Reptiles: Pterosaurs* (1996, Barnes & Noble).

**Figure 8.1:**
Photo of the original specimen of *Pterodactylus antiquus*, first described by Cosimo Collini in 1784 and currently housed in the Bayerische Staatssammlung für Paläontologie und Geologie, Munich, Germany. (Photo courtesy of S. Christopher Bennett; used by permission.)

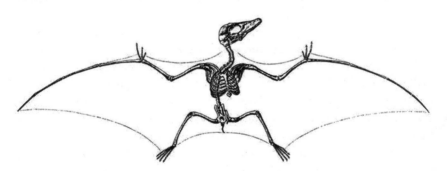

**Figure 8.2:**
Soemmerring's bat-like pterosaur reconstruction.[4]

pterosaur illustration continued to influence the attitudes of both scientists and the general public into the late 20th century.[5,6] Although pterosaur and bat wings both consist of elongated hand and finger bones supporting a flexible membrane, the resemblance is superficial. As we will see in this chapter, pterosaur wings worked rather differently from bat wings.

Over the years since that first *Pterodactylus* discovery, scientists have unearthed a number of fossils that preserve some of the wing membrane. As researchers looked more carefully at those preserved wing membranes, they began to realize that the pterosaur wing membrane was fundamentally different from that of a bat. The bat's membrane is elastic and flexible, and its shape (especially camber) is maintained largely by the bones of the third, fourth, and fifth fingers embedded in the membrane. Pterosaurs, in contrast, have no finger bones to shape the membrane other than the elongated fourth finger that forms the leading edge. Instead, pterosaur wings appear to have a dense array of stiffening fibers in most of the wing that fan out from the wing finger. Researchers are still debating exactly how these fibers worked, but they do agree that the pterosaur wing membrane was not simply a flexible sheet flapping in the breeze like a sail or a flag.

## PTEROSAUR PHYLOGENY: DINOSAUR COUSINS?

We haven't discovered any fossils of proto-pterosaurs in the process of evolving wings, so one way to learn about the early evolution of pterosaurs is to try to figure out what creatures they are most closely related to; we can then use that relationship to infer the characteristics of primitive pterosaurs and their ancestors. On a large scale, pterosaurs are archosaurs. As we saw in Chapter 6, archosaurs include dinosaurs—a classification now containing birds—crocodiles and relatives, and "thecodonts." ("Thecodonts" are no longer considered one lineage but several distinct lineages of primitive archosaurs.) Archosaurs are defined by a characteristic set of openings in the skull, and pterosaurs are clearly archosaurs, but who among the archosaurs are their closest relatives?

### Traditional View

Scientists have traditionally considered pterosaurs to be closely related to dinosaurs. As far back as the first year of the 20th century, scientists such as Harry G. Seeley suggested that pterosaurs shared a close common ancestor with dinosaurs.[7] Just how close has been much debated. For example, German paleontologist Friedrich von Huene published a reconstruction of a small, primitive archosaur called *Scleromochlus taylori* in 1914 and suggested it might be similar to or a relative of the ancestors of pterosaurs.[8] In 1999, Michael Benton took this a step further and suggested the *Scleromochlus* gave rise to both pterosaurs and dinosaurs.[9] Others are not so sure, partly because the *Scleromochlus* fossil is not very detailed and partly because it has short front limbs and long hindlimbs, which seems backward for a proto-pterosaur.[10]

Other researchers have looked for pterosaur ancestors among more primitive archosaurs like *Euparkeria*, or even among reptiles more primitive than archosaurs.[11] Nevertheless, Kevin Padian and Jacques Gauthier defined what is closest to the mainstream view by proposing *Lagosuchus* as a pterosaur ancestor.[12,13] *Lagosuchus*\* was a small (30-cm, 12-inch) bipedal archosaur that is probably at or very near the base of the dinosaur lineage (Fig. 8.3). If pterosaurs are also descended from *Lagosuchus*, that would make dinosaurs and pterosaurs sister groups—that is, each other's closest relatives. A close relationship between *Lagosuchus* and pterosaurs also has important implications for the evolution of pterosaur flight, as we will see.

**Not Dinosaur Cousins?**

Although many paleontologists support a sister-group relationship between pterosaurs and dinosaurs, not everyone agrees. Paleontologist Michael Benton initially placed pterosaurs at the very base of the archosaur tree,[14] but later he decided Padian's analysis was more compelling. Pterosaur specialist (and University of Kansas alum) S. Christopher Bennett has been a prominent critic of the pterosaur-dinosaur connection. In a 1996 paper, he presented a phylogeny that showed pterosaurs down near

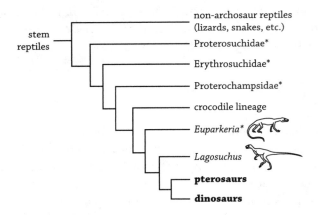

**Figure 8.3:**
Phylogeny showing pterosaurs and dinosaurs as sister groups, and both sharing a close common ancestor with *Lagosuchus*. Asterisks indicate primitive archosaur groups formerly lumped together as "thecodonts." (Redrawn and simplified from Gauthier.)[12]

\* The fossil of *Lagosuchus* is not well preserved. Some paleontologists consider the better-preserved *Marasuchus* either a very close relative or the same species, and *Marasuchus* gives paleontologists much more anatomical information to work on.

the base of the archosaur lineage, among the most primitive archosaurs and well separated from dinosaurs.[15] Although some paleontologists faulted certain technical aspects of his phylogenetic methods, he was not alone. Other studies of archosaur relationships, while focused mainly on more primitive branches, nevertheless place pterosaurs down among those primitive groups rather than at the top with dinosaurs.[16] In a recent, more refined analysis, Bennett changed his methods to answer his critics and still got a similar result: pterosaurs splitting off the archosaur family tree much farther down the trunk than dinosaurs (see Fig. 8.4).[17]

The mainstream view among paleontologists, however, still places pterosaurs as very close relatives of dinosaurs. Numerous studies (but all published before Bennett's most recent work) contain phylogenies showing that close relationship.[18-20] Given the long history of the idea that dinosaurs and pterosaurs are sister groups, plus all those relatively recent studies supporting that view, it remains to be seen whether Bennett's arguments can shift the majority opinion.

Linking pterosaurs evolutionarily to *Euparkeria*, *Scleromochlus*, *Lagosuchus*, or dinosaurs affects how we view the evolution of flight in this group. All these potential precursors have longer hindlimbs than front limbs. Scientists think *Euparkeria* may have been able to run short distances on its hind legs, and *Scleromochlus* may have been partly or largely bipedal. The hind legs of *Scleromochlus* are so long that paleontologists interpret it as a leaper, somewhat in the mold of a wallaby or a lemur. Researchers have used the traits of these potential ancestors as a starting point for theories of flight evolution in pterosaurs.

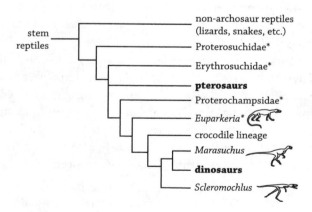

**Figure 8.4:**
Bennett's phylogeny moves pterosaurs to a branch among the primitive archosaurs, well separated from dinosaurs. (*Marasuchus* on this tree corresponds to *Lagosuchus* on the tree in Figure 8.3; asterisks as in Figure 8.3.) Slightly simplified from phylogenies in Bennett.[17]

## THEORIES OF FLIGHT EVOLUTION

The traditional view of flight evolution in pterosaurs is based on a reconstruction of the wing membrane attaching along the hindlimbs. If the hind legs are incorporated into the wings, then running would be difficult if not impossible. Thus, for most of the 20th century, scientists took for granted that pterosaurs, like bats, must have initially evolved wings for gliding. In other words, these scientists assumed that pterosaurs must have evolved flight via some variation on the arboreal (trees-down) mechanism.

In the early 1980s, paleontologist Kevin Padian proposed a radically different scenario that challenged this view.[5,21,22] He suggested that scientists had been misled by thinking that pterosaur wing membranes were necessarily bat-like (harking back to Soemmerring's misconception). He proposed that the wing membranes extended only to pterosaurs' hips, not down the legs, and made the novel suggestion that pterosaurs were bipedal runners. He reasoned that pterosaurs were close relatives of primitive dinosaurs, and primitive dinosaurs were bipedal runners; therefore, primitive proto-pterosaurs would most likely have evolved flight via the cursorial (ground-up) mechanism. Padian called his reconstructions of narrow-winged, bipedally running pterosaurs "birdlike" reconstructions, as opposed to the traditional "bat-like" reconstructions (Fig. 8.5). He and his collaborators presented a new view of pterosaurs as dynamic and agile on the ground as well as in the air,[23] which meshed well with the then-emerging view of dinosaurs as active, warm-blooded, and birdlike. Given this interpretation of pterosaurs, Padian's suggestion of a cursorial origin of flight in pterosaurs seemed natural and logical, and in many ways it mirrored his outspoken advocacy of a cursorial evolution of flight in birds.

Unfortunately for this attractive theory, pterosaur specialists began pointing out that it was not supported by fossil evidence. For one thing, a number of recently described pterosaur fossils that preserved extensive wing membrane seem to show that the wing membrane extended all along the leg to the ankle, so the legs were not free from the wing membranes.[24,25] Moreover, contrary to Padian's reconstruction, both the hip joints and the foot anatomy of pterosaurs seem to be poorly adapted to running.* [26,27] The final nail in the coffin of bipedally running pterosaurs came when

---

\* Padian reconstructed his pterosaurs running on their toes with the heel well off the ground, typical of swift runners like ostriches or antelope. Both foot bones and trackways show, however, that pterosaurs walked with their heels on the ground, like us. This arrangement makes for a fine all-purpose walking and climbing foot, but it is poorly adapted for fast running.

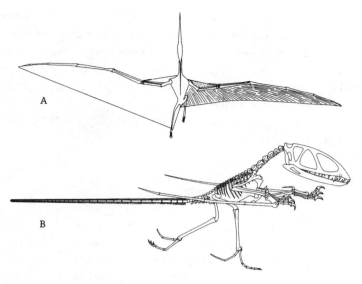

**Figure 8.5:**
Padian's reconstruction of pterosaurs: (A) with narrow wings (right) compared with the traditional broad-winged reconstruction (left); (B) as agile runners on the ground. (A from Padian,[22] used by permission of John Wiley & Sons; B from Padian,[5] used by permission of The Paleontological Society.)

paleontologists described several sets of fossilized tracks that clearly show pterosaurs to have been quadrupedal—walking on all fours—using the wings* and hind legs.[28,29] The current view is well summed up by the title of a 1999 review article, "Pterosaurs: Back to the Traditional Model?":[30] the hind leg was incorporated into the wing membrane, pterosaurs (at least the earlier ones) walked on all fours, and they were not bipedal runners.

If the earliest pterosaurs were quadrupedal walkers rather than bipedal runners, then a cursorial origin for pterosaur flight does not seem logical. A cursorial origin requires bipedal running—first, to free up the forelimbs so they can function aerodynamically, and second, so the animal can run fast enough for the aerodynamics of its forelimbs to matter. Given the widespread agreement that early pterosaurs were quadrupeds, most pterosaur researchers have returned to the view that pterosaurs must have evolved through a gliding stage, based on some variation of the arboreal theory.[31] In this view, the ancestors of pterosaurs must have been arboreal

---

* The handprints in these trackways are quite unlike handprints of any other animals, with impressions of the small fingers at an unusual angle plus a large, odd, trailing impression from the grounded part of the wing finger.

animals that routinely leaped among branches. They would have initially evolved flaps of skin on their forelimbs to help steer or extend leaps (or both). As the flaps enlarged to provide more effective steering, they eventually became extensive enough to generate useful lift, and this would extend leaps and lead to further enlargement. As ancestral pterosaurs became reasonably adept gliders, steering movements would have evolved into rudimentary flapping to extend glides and finally into fully developed flapping for powered flight.

## HOW DID PTEROSAUR WINGS WORK?

Padian's reconstructions, while not ultimately accepted, nevertheless raised useful questions. He argued (correctly) that the pterosaur wing membrane functioned very differently from that of the bat wing, and this argument stimulated much research on the stiffening fibers of the pterosaur wing membrane. As a result, the general consensus is that pterosaur wings, while not as narrow as Padian proposed, were probably narrower than the earlier, traditional, somewhat bat-like depiction.[32] This shape would have made them more aerodynamically efficient and mechanically sophisticated than wings based on the traditional 20th-century view.

A simple, stretchy wing membrane running from the wing finger to the flanks and leg would have flapped like a flag and made an ineffective wing. The stiffening fibers are the key. Some scientists have proposed that the fibers were simply battens like those found on sails to prevent fluttering[24] or that they gave the wing its camber and transmitted flight loads to the wing finger.[23] Bennett proposed in a more sophisticated analysis that the fibers mainly kept the membrane spread from front to back, somewhat like the ribs of an umbrella, while at the same time distributing flight loads away from the tip of the wing finger and more back toward the hand and arm.[32] This arrangement allows the membrane to collapse compactly like a fan when folded, yet retain its fully extended area and camber when flapping. These fibers were only present in the outer three-quarters or so of the wing; the shape and movements of the inner part of the membrane without stiffening fibers was probably controlled mainly by the hind legs.

Imagining pterosaurs walking on the ground may be difficult, but the fossil trackways show that they walked with their legs fairly close together under the hips. They walked with the wings slightly splayed to the side, more or less on the small fingers and the joint at the base of the wing

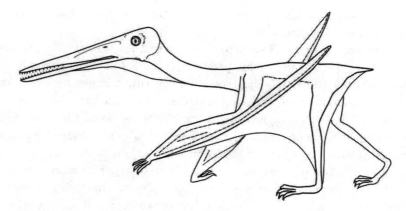

**Figure 8.6:**
A pterosaur walking quadrupedally, based on Bennett's reconstruction,[28] used by permission of Taylor & Francis. (Restoration by S. T.)

finger. The joints between the bones of the wing finger itself were basically immobile and permanently extended so the wing finger would have stuck up and back alongside the animal's flanks as it walked (Fig. 8.6). Because of those immobile wing finger joints, pterosaurs would not have been able to fold the wing compactly like a bird, and this arrangement might have been awkward in tight spaces. On the other hand, the non-wing fingers were strong, well developed, and furnished with strong, curved claws; unlike birds, pterosaurs probably used their forelimbs as much as their hindlimbs for climbing.

## GROWTH IN PTEROSAURS

Pterosaurs were apparently unique among flapping flyers in that they could fly at all body sizes—from juvenile to adult. Consider the other flying animals. Insects, for example, have wings only as adults and adults never molt, so insects don't change their body size once they attain flight. Although birds are not constrained by shedding an exoskeleton, they don't become effective flyers until they are very close to adult body size (mainly because they don't develop flight feathers on their wings until they have achieved most of their growth). Even bats, which possess a more or less complete wing from birth, do not become fully powered flyers until fairly close to adult body size. Indeed, bats are born at a very large body size and they grow so rapidly that when they achieve fully powered flight a few weeks after birth, their wingspan is just slightly shorter than an adult's—their wings are within the range of adult aspect ratios and wing loadings.[33]

In contrast, pterosaurs seem to have hatched at a very small body size and to have spent years reaching adult size. The Solnhofen limestone has produced over 100 fossils of pterosaurs, covering a huge range of body sizes. For many decades, each new specimen was treated as a new species. In a set of detailed comparisons in the 1990s, however, Chris Bennett showed that many of these "species" were actually members of a single species of different ages.[34,35] Based on what appear to be distinct year classes, Bennett suggested that these moderately large pterosaurs spent at least two or three years as slow-growing, independent, flying juveniles before reaching adult size, with adults having wingspans six or eight times that of the smallest juveniles. (*Pterodactylus antiquus* had an adult wingspan of roughly 2 meters or 80 inches so its hatchings might have had wingspans as small as 28 centimeters or 11 inches.) That these juveniles were on their own has been confirmed by recent finds of pterosaur fossils with eggs. The eggs had thin, flexible shells, indicating they were probably buried rather than brooded like bird eggs.[36,37] The small egg size further supports the idea that pterosaurs hatched at a relatively small body size and lived independently for a long juvenile period during which they were presumably capable of flight.[38]

In a nutshell, any given species of pterosaur must have been able to fly over a huge range of body sizes. Flying over such a large range of body sizes is startling not only for its uniqueness among flying animals but because the aerodynamic properties of wings change substantially over exactly this size range. Small wings have markedly lower lift and higher drag than geometrically identical large wings at the extremes of this size range due to the difference in Reynolds number (see Chapter 3, Box 3.1). As a result, flight would have been more energetically costly and less efficient for a very young pterosaur, but it would probably have been able to fly much slower and had more maneuverably than its adult relatives. I am surprised that I can't find any analyses of the consequences of this growth effect. For example, load-carrying and long-distance flight would have been more difficult, but flight in cluttered habitats like forests would have been easier for small, juvenile pterosaurs than for their fully grown parents. These differences would have affected everything related to flight—from migration ability to foraging style—and they deserve a thoughtful analysis. Perhaps the changes in wing properties with size were not as dramatic going from hatchling to adult pterosaurs as going from, say, hummingbirds to hawks, but we won't know until some researcher actually makes the measurements.

## WHAT FOSSILS CANNOT TELL US

Even the best-preserved pterosaur fossils reveal almost nothing about the internal organs of the animal. We thus have no direct evidence for what those organs looked like or how they worked. Were pterosaurs ectothermic like crocodiles, or endothermic like birds, or somewhere in between? Did they have efficient, flow-through lungs like birds or more prosaic tidal lungs like lizards and mammals? The fossils give tantalizing hints, but so far, no unequivocal answers. For example, some pterosaurs had extensively pneumatized (air-filled) bones. Some researchers have taken this as evidence of a birdlike air-sac system and flow-through lungs,[39] but perhaps pterosaurs just evolved skeletal connections with the respiratory system strictly for weight reduction, independent of the lung arrangement. Without information about the pterosaur's internal organs, we have no way of deciding which is more likely.

When fossils are all you have to work with, teasing out anything about the behavior of the living animal is extremely challenging. Some things are fairly obvious: teeth, for example, or—if you are really lucky—stomach contents can reveal dietary preferences. Vertebrate inner ear structures, preserved in skull bones, can show how the animal usually held its head; trackways can tell us how an animal walked or ran. Fossils are mute on many other forms of behavior: how did they find mates? How social were they? Did they build nests or migrate? Did they make sounds? For now, these questions appear to be unanswerable.

### Body Mass

Even estimates of body mass in pterosaurs generate controversy. Body mass is a key element in understanding flight mechanics, so researchers have put a lot of effort into pterosaur body mass estimates. For example, body mass estimates for *Quetzalcoatlus northropi*, the largest known pterosaur, range from 70 kilograms (around 150 pounds) to 540 kilograms (over half a ton!).[40,41] Seventy kilograms seems unrealistically low for an animal that may have stood over 2 meters (7 feet) tall at the shoulder and had the wingspan of a Piper Cub airplane. On the other hand, a half-ton *Quetzalcoatlus* would have been quite incapable of flight, so what use did it have for a wing with a span of over 9 meters (30 feet)? Even the more reasonable sounding mass estimates—200 to 250 kilograms (440 to 550 pounds)—give one pause: could a 250-kilogram pterosaur really have landed slowly enough to avoid injury?[42]

## Heads and Feet

The function of some pterosaur structures remain mysterious. For example, many (perhaps most) pterosaurs had large crests on the top or back of the head (Fig. 8.7). Some of these doubled or tripled the surface area of the head, and some may have been even further extended by soft tissue. Structures this large would inevitably have had significant aerodynamic effects. Paleontologists have suggested that crests were used in flight for steering,[43] and aerodynamic tests have showed that heads with large beaks and crests make very effective rudders.[44] Pterosaurs were unlikely to have used crests as a primary steering mechanism, however: wings are much more efficient and effective for producing turns, and living flying animals steer entirely with wings (a few steer mostly with wings and a little with tails); pterosaurs without crests surely turned using wings just like all other flying animals. Moreover, boats have rudders at the back for a reason: rudders in front are unstable and difficult to keep centered; they are difficult to straighten once turned and very prone to progressively tightening turns once a turn starts. While not the primary steering device, crests could conceivably have been used for fine adjustments to turns or, more likely, for sudden, rapid turns in urgent situations. An intriguing advantage of using a crest to turn is that merely turning the head to look

**Figure 8.7:**
A sample of the variety of pterosaur head crests. (Courtesy of S. T.)

at something would cause the pterosaur to swerve in that direction. If not rudders, what were crests for? Perhaps they were used for some sort of display, to establish dominance or to attract mates. Whatever the function, if crests were not for steering, the function must have been very important because huge crests would have produced significant aerodynamic costs. For example, these large structures would have increased drag and made turning the head to look to the side almost impossible in straight and level flight.

Another pterosaur mystery structure is webbed feet: fossils show that several pterosaur species had them. "Webbed feet?" you say. "That's a no-brainer, they were swimmers." That is also what paleontologists thought. But recently, a study by British paleontologist David Hone and his Canadian colleague Donald Henderson used computer models to show that because they have proportionally huge heads and tiny bodies, pterosaurs would not have been able to float on the surface of the water without tipping over![45] The problem, as contended by Hone and Henderson, is that even though some pterosaurs had long necks, they were not flexible enough to form the sharp S-curve needed to bring the head back over the center of gravity of the body like a floating pelican. With their heads out in front of the body, these pterosaurs could only have floated with the head partly submerged, which would have made breathing a dicey proposition in any conditions other than flat calm. So some pterosaurs may have had webbed feet, but if they used them for swimming and Hone and Henderson are correct, they did not swim like ducks or geese.

### No Molecular Phylogenies

Another limitation of fossils, particularly for building phylogenetic trees, is that scientists cannot get genes from fossils as old as pterosaurs (*Jurassic Park* notwithstanding); the DNA in the genes breaks down far too rapidly. That means researchers trying to reconstruct phylogenies of pterosaurs and their relatives can't tap that important source of phylogenetic information. Instead, they can only use anatomy—mostly limited to bone anatomy, to boot—to build their trees. Some might argue that being limited to anatomical information is not that big a problem; until about two decades ago, that was what all phylogenies were based on, and in many cases adding genes to phylogenetic trees did not change them all that much. Sometimes, however, molecular (genetic) evidence does matter a lot, as we saw for bats (Chapter 7). The argument about the placement of pterosaurs within the archosaur lineage is a case in point.

Lacking molecular data, paleontologists only have anatomical clues to try to tease out whether the apparent similarities of the hindlimbs of pterosaurs and dinosaurs are due to inheritance from a common ancestor, convergent evolutionary responses to similar selection pressures, or sheer coincidence. While most recent studies assume common ancestry, Bennett's study provides the strongest argument for convergence or coincidence—that is, against a close dinosaur-pterosaur relationship.[17] Ironically, Bennett's study borrowed techniques for comparing molecular and anatomical phylogenies and instead used them to compare different categories of anatomical data.

**REASON FOR OPTIMISM?**

Rather than ending our look at pterosaurs on a pessimistic note, we have good reason for a more hopeful outlook. A recent review pointed out that approximately a third of all known pterosaur species have been discovered since 2000.[38] In other words, thanks to huge numbers of new pterosaur fossil finds, mostly in China but also in Brazil, scientists have described half as many new pterosaur species in the last decade or two as for the preceding two centuries! Many of these new fossils are from sites that sometimes preserve soft tissue. The bonanza of new fossils has already helped refine our notions of pterosaurs, and if it continues, at least some of the currently unanswerable questions may finally get resolved.

**REFERENCES**

1. C. A. Collini (1784) *Acta Academiae Theodoro-Palatinae, Mannheim, Pars Physica*.
2. G. Cuvier (1801) *Journal de Physique, de Chimie et d'Histoire Naturelle*.
3. G. Cuvier (1809) *Annales du Muséum national d'Histoire naturelle, Paris*.
4. S. T.v. Soemmerring (1817) *Denkschriften der koniglichen bayerischen Akademie der Wissenschaften München, mathematisch-physikalische Classe*.
5. K. Padian (1983) *Paleobiology*.
6. K. Padian (1991) in *Biomechanics in Evolution*.
7. H. G. Seeley (1901) *Dragons of the Air, an Account of Extinct Flying Reptiles*.
8. F. v. Huene (1914) *Geologische und Paläontologische Abhandlungen NF*.
9. M. J. Benton (1999) *Philosophical Transactions of the Royal Society of London Series B-Biological Sciences*.
10. P. Wellnhofer (1996) *The Illustrated Encyclopedia of Prehistoric Flying Reptiles: Pterosaurs*.
11. R. Wild (1984) *Naturwissenschaften*.
12. J. A. Gauthier (1986) in *The Origin of Birds and the Evolution of Flight*.
13. K. Padian (1984) in *Third Symposium on Mesozoic Terrestrial Ecosystems*.

14. M. J. Benton (1985) *Zoological Journal of the Linnean Society.*
15. S. C. Bennett (1996) *Zoological Journal of the Linnean Society.*
16. S. Renesto and G. Binelli (2006) *Rivista Italiana Di Paleontologia E Stratigrafia.*
17. S. C. Bennett (2013) *Historical Biology.*
18. S. L. Brusatte, M. J. Benton, J. B. Desojo, et al. (2010) *Journal of Systematic Palaeontology.*
19. D. W. E. Hone and M. J. Benton (2007) *Journal of Systematic Palaeontology.*
20. S. J. Nesbitt (2011) *Bulletin of the American Museum of Natural History.*
21. K. Padian (1982) *Sciences-New York.*
22. K. Padian (1985) *Palaeontology.*
23. K. Padian and J. M. V. Rayner (1993) *American Journal of Science.*
24. D. M. Unwin and N. N. Bakhurina (1994) *Nature.*
25. R. A. Elgin, D. W. E. Hone, and E. Frey (2011) *Acta Palaeontologica Polonica.*
26. D. M. Unwin (1987) *Nature.*
27. P. Wellnhofer (1988) *Historical Biology.*
28. S. C. Bennett (1997) *Journal of Vertebrate Paleontology.*
29. D. M. Unwin (1997) *Lethaia.*
30. D. M. Unwin (1999) *Trends in Ecology & Evolution.*
31. S. C. Bennett (1997) *Historical Biology.*
32. S. C. Bennett (2000) *Historical Biology.*
33. L. V. Powers, S. C. Kandarian, and T. H. Kunz (1991) *Journal of Comparative Physiology a-Sensory Neural and Behavioral Physiology.*
34. S. C. Bennett (1995) *Journal of Paleontology.*
35. S. C. Bennett (1996) *Journal of Vertebrate Paleontology.*
36. J. C. Lü, D. M. Unwin, D. C. Deeming, et al. (2011) *Science.*
37. D. M. Unwin and D. C. Deeming (2008) *Zitteliana Reihe B.*
38. D. W. E. Hone (2012) *Acta Geologica Sinica-English Edition.*
39. R. J. Butler, P. M. Barrett, and D. J. Gower (2009) *Biology Letters.*
40. S. Chatterjee and R. J. Templin (2004) *Geological Society of America Special Papers.*
41. D. M. Henderson (2010) *Journal of Vertebrate Paleontology.*
42. M. P. Witton and M. B. Habib (2010) *PLoS One.*
43. W. B. Heptonstall (1971) *Scottish Journal of Geology.*
44. H. R. Jex (2000) *Making Pterodactyls Fly (QN Story).*
45. D. W. E. Hone and D. M. Henderson (2014) *Palaeogeography, Palaeoclimatology, Palaeoecology.*

CHAPTER 9

# Pedestrians Descended from Flyers

*Loss of Flight*

I have spent much of this book discussing the advantages of flapping flight, yet lots and lots of flightless animals are descendants of flying species. We know from their evolutionary relationships that animals like lice, bedbugs, penguins, and ostriches all had ancestors that were fully capable of powered flight, yet they themselves are completely flightless. These secondarily flightless animals demonstrate that species can find themselves in situations where the costs of flight outweigh the benefits. In those situations, natural selection will favor the evolutionary loss of flight.

**ANCESTRY CONSTRAINS LOSS OF FLIGHT**

The main lineages of flying animals differ dramatically in their propensity to abandon flight. Insects include several major lineages (such as fleas and lice) and hundreds, maybe thousands, of minor ones that are secondarily flightless. Many bird lineages also contain flightless members. Aside from large, flightless runners like ostriches and emus, scientists have catalogued flightless ducks, geese, ibises, rails, and even a flightless parrot. The extinct dodos and the threatened kiwis are entirely flightless. In contrast, all known bats can fly, and all but a tiny handful of bat species avoid even landing on the ground. Similarly, to the best of our knowledge, all pterosaurs were flyers. Although the fossil record is incomplete, we have yet to find a pterosaur with greatly shortened or non-wing-like front limbs.

Does the presence of many secondarily flightless insects and birds, but no secondarily flightless bats or pterosaurs, give us any useful insights? The difference is probably due to the degree of independence from the wing structure of some walking legs. Insects' wings are completely separate from their legs, so alone among powered flyers, insects did not give up any legs to achieve flight. Although birds did lose the front limbs in order to evolve flight, the hindlimbs always remained separate; once they developed powered flight, the hind legs were available to evolve separately.* Birds could thus evolve hind legs specialized for all sorts of terrestrial or aquatic locomotion without impairing their ability to fly. This separation of flying locomotion from non-flying locomotion means that birds and insects could evolve such competent non-flying locomotion that under the right conditions they might actually benefit from giving up flight.

In contrast, bats (and presumably pterosaurs) use a flight mechanism that incorporates the hind legs into the wing structure. As a result, natural selection cannot easily act to improve the hindlimbs' non-flight locomotion because such modifications would come at the expense of the animal's flight ability: becoming a slow, clumsy runner is unlikely to be an advantage if it also means becoming a weak, clumsy flyer. This incorporation of the hindlimbs into the wing seems to have committed bats entirely to flight as their primary locomotion. As well as we can tell, the same also applies to pterosaurs.

## LESS-THAN-TOTAL FLIGHTLESSNESS

Loss of flight does not have to be permanent or species-wide. Many birds, especially swimmers, lose the ability to fly when they undergo molt (periodic feather replacement). Most birds molt wing feathers gradually so their ability to fly is only slightly impaired at any one time, but a number of birds seek out refuges with adequate food and few predators and molt all at once. For example, some mergansers (a kind of duck) spend up to 30 days flightless during their molt.[1] Similarly, many species of aphids (tiny plant-feeding insects) are wingless at the beginning of the growing

---

 * While birds may well have gone through a four-winged gliding stage (Chapter 7), the hindlimb was never structurally incorporated into the front wing. As the front limb became more specialized for flapping, the hindlimb appears to have reverted back to a more terrestrial function. Because the hind "wing" never lost separate toes, claws, or leg muscles, this would not have been a dramatic transition.

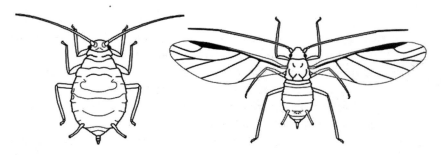

**Figure 9.1:**
Adult female wingless (left) and winged (right) apple aphids of the same species, *Aphis pomi*. (Redrawn by S. T. from Carpenter.)[2]

season when their host plants are also growing and becoming more abundant. Later in the season, crowding causes the same aphid species to produce winged adults that can disperse to seek new populations of the host plant or to head south for the winter (Fig. 9.1).[2,3]

Many species of insects include both flying and non-flying individuals simultaneously. Several species of crickets and many moth species include both long-winged (flying) and short-winged or wingless (non-flying) individuals.[4,5] Sometimes the difference is by gender, with the males flying and the females being flightless; in other species, either sex can become flightless. Flightless females often have bigger ovaries and lay more eggs, so species with both flying and flightless forms seem to exist in a balance where some situations favor laying more eggs but other situations favor more dispersal ability.

Social insects like ants and termites also include flying and flightless individuals in the same species. In these species, wingless, sterile workers make up the vast majority of the colony's members. Only reproductive individuals have wings. These winged individuals provide the main dispersal route for the species. They leave the colony, mate, and disperse to new locations where the survivors found new colonies. In fact, once the mated queen finds a good location to start a new nest, she promptly sheds her wings and never flies again. I am tempted to suggest that the subterranean ways of these social insects promoted wing loss in the workers. Indeed, many other burrowing and wood-boring insects (wood roaches, numerous beetles) have also lost their wings, yet many additional burrowing and wood-boring species have retained their wings. So burrowing may or may not contribute to flightlessness, depending on the specific circumstance.

Often, flightlessness is fixed and permanent. Ostriches are a classic example. Because of their large size, specializations for running, and relatively small, fluffy wings,* all ostriches are permanently pedestrian. Similarly, penguins have become so specialized for underwater swimming that their flipper-like wings, albeit powerful, are far too small for aerial powered flight. Among insects, many such as fleas and lice that live as external parasites (ectoparasites) on vertebrates are completely flightless, evidently dispersing along with their hosts rather than flying.

## WHEN AND WHERE NOT TO FLY

Charles Darwin may have been the first to suggest that oceanic islands might promote flightlessness.[6] Flyers on islands may not have anywhere much to go for dispersal. Moreover, strong oceanic winds and especially storms might blow a flyer so far from the island that it could not find its way back. While no problem for a seabird, getting blown far off an island could be fatal for a wren or a June beetle. Scientists have long argued about whether islands really promote flightlessness, possibly because the answer depends on what animal a scientist studies. Although Pacific islands do have many flightless insects (which is what originally caught Darwin's interest),[7-9] evolutionary biologist Derek Roff has analyzed the distribution of flightlessness in insects and concluded that flightlessness is no more common among island insects than among mainland insects.[5]

### Land Birds

Birds, however, are a different story. Flightless island bird species abound, from the extinct dodos, moas, Hawaiian geese, and Jamaican ibises to living Galapagos cormorants and New Zealand's kiwis and flightless parrots. Many of these are medium-sized birds that are large enough that they really can get anywhere on an island by walking, and on many islands, they have no terrestrial predators. Perhaps the best-known examples are the flightless rails of the Pacific islands. Rails are ground-feeding, somewhat chicken-like birds that are widely distributed around the world. On continents, they normally can fly. On many Pacific islands, however,

---

* Ostriches can have a wingspan of over 1.5 meters (5 feet), which seems pretty large until you realize that they stand 1.8 to 2.7 meters (6 to 9) feet tall and weigh over 90 kilograms (200 pounds). The extinct giant teratorn, which almost certainly could fly, only weighed three-quarters as much as an ostrich but had wingspans well over 6.7 meters (22 feet).

they cannot. Phylogenetic studies show that once rails colonize a predator-free island, they lose flight fairly quickly in evolutionary terms—120,000 to 500,000 years.[10] Environmental physiologist Brian McNab showed that not only do flightless island birds escape the high energy cost of flying but their resting energy consumption is lower than that of their flying relatives. That difference allows the flightless birds to survive on less food, or to produce more eggs on the same food, than their flying relatives.[11]

### Waterfowl

Several different lineages of aquatic birds have become flightless. Penguins are the most obvious example, but they are far from alone. Other lineages include flightless cormorants, steamer ducks, New Zealand teals, the extinct flightless ducks and geese of Hawaii, and the recently extinct Great Auk. Diving birds actually benefit from being denser or heavier for their size, a condition that obviously conflicts with flight. Shorter, stubbier wings are more effective for swimming but less effective for flying. These birds can use blubber (a fat layer) instead of feathers for insulation; such fat would be prohibitively heavy for a flyer but it has several advantages for a diver. For example, air trapped in the feathers of a diver compresses with increasing depth and changes the bird's buoyancy whereas blubber doesn't compress so it does not affect the bird's buoyancy as the bird dives deeper.

The Great Auk, sadly driven to extinction just a couple hundred years ago, is a particularly revealing example. The auks are a family of birds (Alcidae) that both fly in air and swim underwater with their wings. Their wings are thus a compromise, small for flight but big for swimming. Curiously, the small auks like dovekies and least auklets (18 to 20 centimeters or 7 to 8 inches long) are the strongest, most agile flyers and the weakest, slowest swimmers, whereas the large ones like murres (44 centimeters or 17 inches long) are powerful swimmers but marginal flyers. Murres appear awkward when landing—sometimes requiring several tries and still sprawling clumsily—and they require such a long takeoff run on the water surface that they are unable to take off in rough water or without a headwind. The Great Auk was the logical extreme: twice as long and five times as heavy as a murre, it was apparently a very fast, capable, penguin-like swimmer, but it was totally flightless (see Box 9.1). Evidently, the benefits of large size for swimming won out over the benefits of marginal flying ability, so as it got bigger, it evolved much shorter, smaller wings for swimming and lost flight. As well as being flightless, Great Auks were somewhat awkward on land and had little fear of humans due to breeding on rocky islands and cliffs without predators. They were thus easy pickings for market hunters who killed them in wholesale batches for their eggs and feathers. By the mid-1800s, they were gone.[13]

When diving birds can breed in environments free from terrestrial predators and can obtain the food they need and get to the places they need to go by swimming, the costs of flight may come to outweigh the benefits, as has happened in all these flightless swimmers.

## Insects

As for insects, if islands don't promote flightlessness, what does? Surprisingly, one of the strongest correlations is with very stable, long-lived, undisturbed habitats. Quite a few species of deep forest moths have flightless females, and these are often found in large, undisturbed, virgin forests.[5] Probably more familiar are the previously mentioned ectoparasitic insects like fleas and lice. Flightlessness makes good sense here: if your home and

---

### Box 9.1: THE ORIGINAL PENGUIN

The term "penguin" originally referred to Great Auks. Linguists disagree on the origin of the term: some derive it from a Welsh phrase for "white head" but others say it comes from a Spanish or Portuguese word that means "fat" in some contexts. Regardless of the origin, British sailors who encountered Great Auks in Newfoundland in the 1500s were calling them "penguins." Indeed, when Carolus Linneus, the father of modern taxonomy, described a Great Auk, he gave it the scientific name *Pinguinus impennis*.

When Sir Francis Drake's ship, the *Golden Hind*, circumnavigated the globe in the late 1500s, his sailors encountered the birds we now call penguins in the Straits of Magellan. They called these Southern Hemisphere birds "penguins," either because of their close resemblance to Great Auks or because the sailors did not realize they were different from Great Auks. (Great Auks and penguins look superficially similar but they are not closely related.) Possibly due to the popular accounts of the *Golden Hind*'s voyage, perhaps aided by the declining Great Auk populations as well as alternative local names like "garefowl," by the 1600s "penguin" began to take on its modern meaning, referring solely to the Southern Hemisphere birds.

your source of food—your host—is mobile, why ever leave? Moreover, if you can't find a suitable mate on your current host, just wait till your host mates. Then you can simply walk onto this new host and look for mates. Flight for such an insect clearly provides little or no benefit, and the cost of leaving (and potentially losing) your host could be quite high. In addition to the major lineages represented by fleas and lice, numerous minor lineages including bedbugs and sheep keds have gone this route. Sheep keds look rather tick-like but they are insects in the order Diptera, so they are actually flightless flies!

Many temperate and high-latitude insects can survive freezing or even sub-freezing temperatures, but they usually do so hunkered down and dormant. Very few insects actively feed and reproduce at freezing temperatures; a handful of insects do, however, and can sometimes be found wandering around on the surface of snow, particularly in alpine regions. None of these cold-loving insects can fly. The problem is physiological. At such low temperatures, muscles simply can't move fast enough to flap the wings effectively. North American examples of such cold-loving insects include the snow scorpionflies in the genus *Boreas* and snow crane flies in the genus *Chionea*,[12] the latter being another example of flightless flies.

## ONCE LOST, GONE FOREVER

If, through natural selection, a species that once could fly evolves loss of flight, can it reverse the process and regain the power of flight? If, by reversal, I mean regaining lost structures such as wing components in their original form and with their original function, evolutionary biologists would overwhelmingly say, "No." Once a structure is lost (assuming the loss is genetically fixed throughout the species), nature has no way to rewind or back up to the pre-loss condition. In principle, natural selection could modify some other structure to function in place of the lost structure. For example, true flies (Diptera) have lost the egg-laying appendage or ovipositor of other insects, but a few flies have modified the back end of the abdomen to function like an ovipositor. When it comes to flight, if a species doesn't fly because its wings are too short or its flight muscles are too small, then under the right conditions, natural selection could conceivably enlarge those structures and allow the species to regain flight (although I have never heard of a case of this happening for a completely flightless species). If, however, major components of the wings or their muscles are completely lost throughout the species, as in the total absence of wings in fleas, then the species will be permanently grounded. (Some researchers think that stick insects may have lost and then reacquired wings based on a molecular phylogeny,[14] but this claim is quite controversial and other scientists have reinterpreted the phylogeny to show no need for wing reacquisition.)[15]

You might say, "Well, couldn't they evolve wings partly or entirely from new structures, the way the first birds or bats did?" and the answer is, "Yes, but only under the right conditions." Considering that such an incipient flyer would be competing with a whole host of already capable flyers, and especially considering that the "right conditions" seem to have occurred only four times in the more than 400-million-year history of land animals, the chances of a secondarily flightless animal "re-evolving" powered flight are, for all practical purposes, zero.

How widespread is loss of flight? Among birds it is quite rare, having occurred in fewer than 60 species out of the 10,000 or so scientifically described species living today. Insects are more challenging to pin down. Flight loss can vary hugely among lineages (for example, 100% of fleas are flightless whereas 0% of dragonflies and damselflies are flightless). Based on the amount of flight loss estimated for most of the major flying lineages,[5] an overall estimate for known insect species is in the 5% to 10% range. Although small on a percentage basis, 5% of nearly a million described insect species is still a pretty big number. In fact, that is more or less the same number as all known vertebrate species (fish, amphibians, reptiles, mammals, and birds) put together.

Flight obviously has its advantages. As we have seen in this chapter, however, animals in some situations may not benefit much from being able to fly and may actually incur evolutionary costs—growing, maintaining, and carrying around wings and flight muscles when flight is counterproductive. Under these conditions, natural selection will actually favor loss of flight.

**REFERENCES**

1. S. R. Craik, J.-P. L. Savard, and R. D. Titman (2009) *Condor*.
2. P. J. Gullan and P. S. Cranston (2010) *The Insects: An Outline of Entomology*.
3. G. H. Carpenter (1913) *The Life-Story of Insects*.
4. D. A. Roff (1984) *Oecologia*.
5. D. A. Roff (1990) *Ecological Monographs*.
6. C. Darwin (1860) *On the Origin of Species*.
7. N. L. Evenhuis (1997) *Bishop Museum Occasional Papers*.
8. M. J. Medeiros (2008) *Zootaxa*.
9. C. A. Tauber, M. J. Tauber, and J. G. Giffin (2007) *European Journal of Entomology*.
10. B. Slikas, S. L. Olson, and R. C. Fleischer (2002) *Journal of Avian Biology*.

11. B. K. McNab (1994) *American Naturalist*.
12. G. W. Byers (1969) *Evolution*.
13. D. N. Nettleship and P. G. H. Evans (1985) in *The Atlantic Alcidae: The Evolution, Distribution and Biology of the Auks Inhabiting the Atlantic Ocean and Adjacent Water Areas*.
14. M. F. Whiting, S. Bradler, and T. Maxwell (2003) *Nature*.
15. J. W. H. Trueman, B. E. Pfeil, S. A. Kelchner, and D. K. Yeates (2004) *Systematic Entomology*.

CHAPTER 10

# Unifying Themes?

As we look for shared features of flight evolution among the Big Four—insects, pterosaurs, birds, and bats—we must keep in mind that each of these groups evolved flight at different times separated by tens or hundreds of million years and under extremely different circumstances. Insects probably evolved flight fairly soon after becoming terrestrial, which would have been before plants had fully colonized land far from water. Pterosaurs arose more than 100 million years later, after early forms of trees had evolved and formed widespread forests. Pterosaurs would not have competed directly with insects because they were so much larger, so these first flying vertebrates also faced no aerial competition. Birds, on the other hand, would have been at risk of competition and possibly predation from pterosaurs when they evolved flight 50 million years or so after pterosaurs. Perhaps birds evolved flight far inland at a time when most pterosaurs were largely coastal, or perhaps the protobirds were enough smaller than pterosaurs that they avoided most competition. Finally, birds and maybe pterosaurs were present when bats first took to the skies. Bats apparently avoided both competition and predation from other large flyers by specializing in nocturnal flight. The ecological and environmental conditions were thus vastly different for each of these lineages as they evolved flight. Nevertheless, they do have a few features in common.

**MAJOR PARALLELS**

The most obvious common feature of the Big Four flyers is flapping, which I have used throughout the book interchangeably with powered flight. All living powered flyers flap their wings for thrust, and everything we know about pterosaurs suggests that they did as well. Flapping is really a consequence of using muscles for power (Chapter 2). Muscles don't allow continuous rotation, so the next-best option is some sort of oscillating or see-saw motion. Given the way muscles work and the way wings work, flapping seems to be the only practical way an animal can produce thrust for flight, at least in a low density medium like air.* Nonetheless, a transition from walking to flapping is not trivial. The right and left legs alternate in typical walking, but they have to move simultaneously in flapping. Interestingly, when my cat climbs a tree in a hurry, he does so with the right and left front legs moving together, alternating with the back legs, which also move together. (This is a variation of the "bounding" gait, which is what locomotion researchers call the hopping gait used by animals like rabbits.) This kind of climbing might have served as an exaptation for the original flapping motion, which is yet another point in favor of arboreal climbers being ancestral to flyers.

Some scientists may disagree, but I see directed aerial descent leading to gliding ("arboreal") evolution of flight as a possible unifying feature. The more researchers look, the more they find some form of aerial maneuverability in climbing animals. Cats always land on their feet, mice and arboreal lizards assume a sky-diving posture and make soft landings, and ants, spiders, and silverfish knocked off a tree branch can steer their fall back to the trunk. These animals all possess the ability to sense and in some way influence their movement through the air during a fall. This ability gives them the sensory and behavioral basis for rudimentary gliding without any overt aerodynamic structures. Full-fledged gliding would only require incremental enlargement of aerodynamic surfaces, which in turn explains why so many kinds of arboreal animals have evolved gliding.

The preponderance of evidence now suggests that all the powered flyers may have started by gliding from elevated perches. Most researchers agree that bats and pterosaurs originated flight from the trees down. Some may not yet accept a trees-down origin for bird flight, although I think the evidence from many recently described feathered dinosaurs is quite compelling. Finally, Yanoviak and colleagues developed the very concept of directed aerial descent based on insects.[1,2] Although the researchers are hampered by lack of fossil evidence, their theory is so physically and

* In air, a flyer's wings need to support the animal's weight as well as produce thrust. In water, a swimmer's weight is mostly supported by the water due to the animal's buoyancy; therefore it only needs to produce thrust, not lift. Aquatic animals can thus make use of intermittent, thrust-only mechanisms like squids' use of intermittent jetting. Nevertheless, the lift mechanism works just as well under water, and in addition to sea lion, penguin, and sea turtle flippers that are obviously underwater flapping wings, a fair number of fish tails—tunas, swordfish, sharks—operate as wings flapping sideways instead of up and down.

biologically reasonable that I would be surprised if directed aerial descent did not play at least some role in the evolution of flight in insects.

Any animal that flies, whether glider or flapper, must be able to sense objects at a distance. Flight speeds demand that a flyer be able to detect obstacles and landing sites long before they are close enough to touch and with much more accuracy than possible with smell. Flying animals thus tend to have acute vision. All flyers tend to have very good vision compared to non-flyers, and birds of prey may have the sharpest vision in the animal kingdom.[3] Bats are the exception that proves the rule. Bats actually have above average vision for their size, but optical physics limits the ability of small eyes to see detail in very dim light. So bats have evolved echolocation to replace vision in the dark. Although echolocation can be amazingly acute, it is a fairly short-range system, so bats revert back to using vision whenever enough light is available.[4]

As a rule, modern flying animals are extremely maneuverable. This maneuverability is a consequence of having little built-in or passive stability (although one study suggests that the flapping motion itself might have some stabilizing effect).[5] The lack of passive stability means that a flying animal can change directions quickly and easily. The flip side of this is that if a flapping flyer intends to fly in a straight line, the animal will need to make continuous, small, rapid course corrections to actively stabilize its course. Some primitive flyers, such as *Archaeopteryx* with its long tail, seem to have had considerable passive stability, but all modern flyers (as well as the last main pterosaur lineage, the pterodactyloids) have evolved toward high maneuverability at the expense of stability. This was surely a response to predation because greater maneuverability makes a flyer better at evading predators. Flying animals do cover a range of maneuverability, with smaller, slower flyers tending to be more maneuverable and larger, faster flyers less so. Even large, relatively ungainly flying animals like swans or cranes are, however, still vastly more agile than even the most maneuverable fighter jet or competition aerobatic airplane.

Gaining that maneuverability is not just a matter of discarding stabilizing structures. The animal needs to replace stabilizing structures with active course-correcting responses, and this means modifying the nervous system to continuously detect and correct for small deviations, typically using reflexes. So along with acute vision, flying animals all seem to have evolved modifications to the nervous system to actively stabilize flight, giving them extreme maneuverability.

Another theme, at least of the vertebrate flyers but less so for insects, is that the specialized demands of flapping flight have greatly limited the structural variation and diversity of flyers. As one ornithologist put it, under their feathers, birds all look remarkably similar.[6] Of course, within birds, or within bats, some variation does exist. But if you think of the variation in body form between a weasel and a domestic cow, or between a rabbit and an armadillo and an elephant, flying vertebrates just don't show that much anatomical diversity. So much of their anatomy is specialized for flapping flight that these specializations greatly restrict major modifications and variations. Indeed, this structural uniformity posed a significant problem for researchers using anatomical traits to develop the earliest bird phylogenies. Once gene-based phylogenies became available, researchers realized that many of the structural characteristics that seemed to unite various groups were actually convergences driven by the physical constraints of flapping flight (see Box 10.1. Convergence and Homology, Trash or Treasure?).

Insects are freed from this constraint partly by their size and partly by their modular body plan. Streamlining provides very little benefit for very small animals (Chapter 3), which frees insects to adopt body shapes that

---

*Box 10.1:* **CONVERGENCE AND HOMOLOGY, TRASH OR TREASURE?**

As we compare animal flight across the Big Four powered flyers, two key concepts surface repeatedly: convergence and homology. When scientists look at two animals with similar structures that do the same thing, they immediately want to know if the two structures evolved independently due to similar selection pressures and constraints—convergence—or whether they are similar because they were inherited in similar form from a common ancestor—homology. Homologies thus arise from ancestry whereas convergences arise when different species face similar constraints.

The constraints that lead to convergence are often physical. For example, aerodynamics dictates the properties of effective wings, so functional wings are normally big, flat, slightly cambered surfaces. Similarly, flyers like vertebrates that operate at higher Reynolds numbers benefit greatly from streamlining. If we ignore the heads and legs of birds, for instance, their bodies are all strikingly similar, that is, compact and streamlined.

Physical constraints may be powerful, but animals are also heavily constrained by their ancestry, so-called historical constraints. Evolution can only work with the limited selection of building blocks present in a lineage's ancestors. For example, animal flyers cannot evolve a propeller for thrust, because their muscles (and their ancestors' muscles) do not

*Box 10.1: Continued*

allow a continuously rotating body part. For this reason, flying animals are constrained to flapping their wings for thrust. The same applies to anatomical properties. Bats are a nice illustration: the ancestors of bats did not have feathers, so their wings make do with skin and bones instead.

Evolution by natural selection has no mechanism to produce massive changes in structure or material composition in a single step. A jumping animal cannot suddenly "decide" to grow full-sized wings the way Frenchman Gabriel Poulain could bolt a pair of wings to a bicycle in the 1920s. Animals can evolve such large structures only gradually because even extremely rapid evolution can still take on the order of a million generations. Similarly, animals have no way to make a jump from the structural materials that compose their ancestors' bodies to totally new and unrelated materials. No matter how much benefit they might gain, no insect can evolve a graphite-epoxy exoskeleton, no bat can evolve titanium bones. Animals are limited to incremental changes in structures built from a limited selection of materials present in their ancestors.

While convergence and homology are both important in evolution, a biologist's perspective on these concepts is shaped by the kinds of questions he or she is exploring. Functional biologists, such as those in my own field of biomechanics, are often most interested in why different animals work in similar ways. Why are the wings of both insects and birds usually cambered? Why are the flapping patterns of all flying animals so similar? Understanding aerodynamics helps us to understand convergent wing structures and vice versa. Without evolutionary convergences, we don't have such comparisons, and homologies yield few functional insights.

In contrast, systematists trying to reconstruct evolutionary history use homologies as the main basis for building phylogenetic trees. Convergences add confusion. If a systematist inadvertently treats a convergence as a shared, derived characteristic, the resulting tree will be less accurate, implying shared ancestry where none exists. So to systematists, homologies are valuable, whereas convergences need to be recognized and eliminated. Recognizing convergent structures can be challenging, but building trees with genetic as well as anatomical traits has greatly improved the process.

Both functional biologists and systematists want to be able to recognize convergences and to be able to tell them from homologies. The difference is that evolutionary convergences are valuable and informative to a functional biologist but detrimental and a source of error to a systematist.[7]

would be too draggy for larger animals. Also, the flight mechanism is entirely contained in one of the three body regions, the thorax, leaving the head and abdomen fairly free from flight-related limits. Flight has thus constrained the thorax to some extent, but this has been a fairly minor restriction on insect body form; this relaxed constraint allowed insects to evolve much more anatomical variety than any lineage of flying vertebrates.

## DIFFERENCES

Among the Big Four, insects are in a class by themselves. Where flying insects probably diverge most from the flying vertebrates is in the origin of the protowings. Whatever structure the earliest insect wing evolved from, it was not a leg. Flying insects still have the same number of legs as primitively flightless insects like silverfish. In contrast, natural selection converted the front legs of birds into wings while leaving the back pair as fully independent legs. Bats and pterosaurs took this a step further, incorporating the hindlimb into the wing structure and thus limiting their terrestrial versatility.

Insects also diverge from vertebrate flyers in size. The largest flying insects—cicadas, Goliath beetles, big dragonflies, and moths—do overlap the size range of hummingbirds, but these animals represent the extremes. The vast majority of flying insects are much smaller than hummingbirds, whereas the vast majority of birds and bats are larger than hummingbirds. For a flyer, size matters. Tiny wings are inefficient, making long-distance flight ineffective. Tiny flyers, however, have high surface-to-volume ratios making vertical takeoff, maneuvering, and hovering easy, and they have low inertia so collisions cause bounces, not injuries. Large flyers face the opposite situation. Their more-efficient wings make long-distance flight economical, but only the smallest can take off vertically or hover, many require long takeoff runs, and collisions with solid objects can be deadly. Within vertebrates, birds probably cover the widest size range, bats tend toward the small end, pterosaurs tend toward the large end, but they all overlap extensively.

## STILL EVOLVING?

Throughout this book we have been looking at major evolutionary events, and some readers may wonder if these animal groups are still evolving. The answer is an emphatic "yes."

Natural selection is still acting, but the process is usually far too slow for humans to observe directly. The loss of flight by rails on islands in 100,000 to 500,000 years (Chapter 9) is extremely rapid in evolutionary terms and barely an eyeblink to a paleontologist. Yet that is also dozens of times longer than the entire span of recorded human history, which shows how unlikely we are to actually witness such significant evolutionary events. The best we can do is look at patterns of relatedness and past speciation (in other words, phylogenies) and use those to decipher recent evolutionary events. Such studies tell us that a good bit of speciation and anatomical modification has occurred among flyers since modern humans arose a million or so years ago. If we could look back in another 500,000 or one million years, we would undoubtedly see that some of the groups that seem to us to be one species of insect or bird are actually undergoing rapid change and may be in the process of forming new species.

**UNANSWERED QUESTIONS**

Although tens of thousands (if not millions) of pages in scientific journals have been devoted to the evolution of animal flight, we still face several major unanswered questions. Our ignorance is due to those pesky gaps in the fossil record. Until some of those gaps are filled in, we can place boundaries around what is physically or biologically possible, but we can never conclusively answer certain questions.

Surely the most vexing unanswered questions are when and from what precursors wings evolved in insects, pterosaurs, and bats. Thanks to those abundant new Chinese fossils, we are fairly close to answering those questions for birds. The frustrating gaps, however, mean that we can't really pin down when wings evolved in the other three groups. Primitive, flightless insects put in a brief appearance very early, followed by a gap of tens of millions of years, after which fully winged insects appear. As for bats and pterosaurs, since we have no transition fossils, we cannot even say with confidence who their direct ancestors were. We cannot tell whether these groups were flying for an enormously long time or just a brief few million years before the lifetime of our oldest fossils.

Similarly, without transition fossils or obvious pre-flight ancestors, the exact function of the protowings—the structures that evolved directly into wings—remains a mystery for insects, pterosaurs, and bats. Obviously pterosaur and bat wings started out as front limbs, but what were the ancestral animals doing with those limbs? Probably not running, but most likely climbing and eventually gliding. Without fossil evidence, however, we can't escape the possibility that they were doing something entirely unexpected. And as for insects, the situation is even worse because researchers really don't know what the protowing was, let alone what it was used for before flight.

We have a pretty good handle on the ancestry of flying insects and birds, but the direct ancestry of pterosaurs and bats remains an open question. Bats are so distantly related to other living mammals that these relationships are not that much help for figuring out their immediate ancestors. In contrast, the long association of pterosaurs with dinosaurs appeared to make their ancestry fairly obvious, but that relationship seems to be unraveling. If pterosaurs are not the sister group of dinosaurs, then researchers need to do a lot more work to figure out who pterosaurs evolved from.

Scientists have a long history of debating several "which came first, chicken or egg?" questions about the order in which some flight-related specializations evolved. For example, did flight drive the evolution of feathers, or did feathers evolve first and then become co-opted for flight? After many decades of discussion, the recent fossil discoveries in China have pretty much answered this one: feathers evolved well before flight. In contrast, several other sequence questions remain unanswered. A few of the most important: Did insect wings evolve from a structure that already had a joint, or did the protowings first evolve their aerodynamic function and then evolve a new joint to allow flapping? Did theropod dinosaurs evolve one-way lungs before birds arose, or did birds evolve flight first and then evolve one-way lungs? Ditto for endothermy. Did bats evolve echolocation before or after evolving powered flight? Either new fossils or new ways of analyzing old fossils will be needed to answer these questions.

All of these unanswered questions can be discouraging, but I am optimistic. Scientists using the bounded-ignorance approach, especially when based on biomechanics and functional morphology, have produced fresh insights and innovative new theories. Modern phylogenetic methods have greatly refined our understanding of bird evolution and could possibly do the same for other groups. Thanks also to the treasure trove of wonderful fossils described over the last decade from China, some of the major questions about flight evolution in birds have been definitely or partly answered. While those fossil beds are too young to help us much with flight origins in pterosaurs or insects, they do show how a new source of fossils can revolutionize our view of the evolution of flight in one lineage. Perhaps sources of fossils that can do the same for other animal groups await discovery by a new generation of researchers.

**REFERENCES**

1. S. P. Yanoviak, R. Dudley, and M. Kaspari (2005) *Nature*.
2. S. P. Yanoviak, M. Kaspari, and R. Dudley (2009) *Biology Letters*.
3. R. Fox, S. W. Lehmkuhle, and D. H. Westendorf (1976) *Science*.
4. J. D. Altringham (2011) *Bats: From Evolution to Conservation*.
5. T. L. Hedrick (2011) *Journal of Experimental Biology*.
6. A. S. King and D. Z. King (1979) in *Form and Function in Birds*.
7. E. O. Wiley and B. S. Lieberman (2011). *Phylogenetics: The Theory of Phylogenetic Systematics*.

# BIBLIOGRAPHY

Abzug, M. J. and E. E. Larrabee (1997). *Airplane Stability and Control: A History of the Technologies That Made Aviation Possible.* Cambridge University Press, Cambridge. 373 pp.

Alexander, D. E. (1986). Wind tunnel studies of turns by flying dragonflies. *Journal of Experimental Biology* **122**: 81–98.

Alexander, D. E. (2002). *Nature's Flyers: Birds, Insects, and the Biomechanics of Flight.* Johns Hopkins University Press, Baltimore, Maryland. 358 pp.

Alexander, D. E. (2009). *Why Don't Jumbo Jets Flap Their Wings? Flying Animals, Flying Machines, and How They Are Different.* Rutgers University Press, New Brunswick, New Jersey. 278 pp.

Alexander, R. D. and W. L. Brown (1963). Mating behavior and the origin of insect wings. *Occasional Papers of the Museum of Zoology of the University of Michigan* **628**: 1–19.

Altringham, J. D. (1996). *Bats: Biology and Behaviour.* Oxford University Press, Oxford. 262 pp.

Altringham, J. D. (2011). *Bats: From Evolution to Conservation.* Oxford University Press, New York. 324 pp.

Anderson, J. D. (2007). *Fundamentals of Aerodynamics.* McGraw-Hill Higher Education, Boston. 1008 pp.

Bailey, W., J. Slightom and M. Goodman (1992). Rejection of the "flying primate" hypothesis by phylogenetic evidence from the epsilon-globin gene. *Science* **256**: 86–89.

Baker, R. J., M. J. Novacek and N. B. Simmons (1991). On the monophyly of bats. *Systematic Zoology* **40**: 216–231.

Bakker, R. T. (1972). Anatomical and ecological evidence of endothermy in dinosaurs. *Nature* **238**: 81–85.

Beck, R. M. D., O. R. P. Bininda-Emonds, M. Cardillo, F.-G. R. Liu, and A. Purvis (2006). A higher-level MRP supertree of placental mammals. *BMC Evolutionary Biology* **6**: 93.

Beckemeyer, R. J. (2000). The Permian insect fossils of Elmo, Kansas. *Kansas School Naturalist* **46**: 1–15.

Bennett, S. C. (1995). A statistical study of *Rhamphorhynchus* from the southern limestone of Germany: Year-classes of a single large species. *Journal of Paleontology* **69**: 569–580.

Bennett, S. C. (1996). The phylogenetic position of the Pterosauria within the Archosauromorpha. *Zoological Journal of the Linnean Society* **118**: 261–308.

Bennett, S. C. (1996). Year-classes of pterosaurs from the Solnhofen limestone of Germany: Taxonomic and systematic implications. *Journal of Vertebrate Paleontology* **16**: 432–444.

Bennett, S. C. (1997). The arboreal leaping theory of the origin of pterosaur flight. *Historical Biology* **12**: 265–290.

Bennett, S. C. (1997). Terrestrial locomotion of pterosaurs: A reconstruction based on *Pteraichnus* trackways. *Journal of Vertebrate Paleontology* **17**: 104–113.

Bennett, S. C. (2000). Pterosaur flight: The role of actinofibrils in wing function. *Historical Biology* **14**: 255–284.

Bennett, S. C. (2013). The phylogenetic position of the Pterosauria within the Archosauromorpha re-examined. *Historical Biology* **25**: 545–563.

Bennett, A. F. and J. A. Ruben (1979). Endothermy and activity in vertebrates. *Science* **206**: 649–654.

Benson, R. B. J., R. J. Butler, M. T. Carrano and P. M. O'Connor (2012). Air-filled postcranial bones in theropod dinosaurs: Physiological implications and the reptile-bird transition. *Biological Reviews* **87**: 168–193.

Benton, M. J. (1985). Classification and phylogeny of the diapsid reptiles. *Zoological Journal of the Linnean Society* **84**: 97–164.

Benton, M. J. (1999). *Scleromochlus taylori* and the origin of dinosaurs and pterosaurs. *Philosophical Transactions of the Royal Society of London Series B-Biological Sciences* **354**: 1423–1446.

Bertin, J. J. and M. L. Smith (1979). *Aerodynamics for Engineers*. Prentice-Hall, Englewood Cliffs, New Jersey. 410 pp.

Béthoux, O. and D. E. G. Briggs (2008). How *Gerarus* lost its head: Stem-group Orthoptera and Paraneoptera revisited. *Systematic Entomology* **33**: 529–547.

Bishop, K. L. (2008). The evolution of flight in bats: Narrowing the field of plausible hypotheses. *Quarterly Review of Biology* **83**: 153–169.

Boag, D. and M. Alexander (1986). *The Atlantic Puffin*. Poole (for Blandford Press, Australia), New York. 128 pp.

Bock, W. J. (1965). The role of adaptive mechanisms in the origin of higher levels of organization. *Systematic Zoology* **14**: 272–287.

Bock, W. J. (1969). The origin and radiation of birds. *Annals of the New York Academy of Sciences* **167**: 147–155.

Brusatte, S. L., M. J. Benton, J. B. Desojo and M. C. Langer (2010). The higher-level phylogeny of Archosauria (Tetrapoda: Diapsida). *Journal of Systematic Palaeontology* **8**: 3–47.

Bryant, H. N. and A. P. Russell (1993). The occurrence of clavicles within Dinosauria: Implications for the homology of the avian furcula and the utility of negative evidence. *Journal of Vertebrate Paleontology* **13**: 171–184.

Bundle, M. W. and K. P. Dial (2003). Mechanics of wing-assisted incline running (WAIR). *Journal of Experimental Biology* **206**: 4553–4564.

Burgers, P. and L. M. Chiappe (1999). The wing of *Archaeopteryx* as a primary thrust generator. *Nature* **399**: 60–62.

Butler, R. J., P. M. Barrett and D. J. Gower (2009). Postcranial skeletal pneumaticity and air-sacs in the earliest pterosaurs. *Biology Letters* **5**: 557–560.

Butler, R. J., P. M. Barrett and D. J. Gower (2012). Reassessment of the evidence for postcranial skeletal pneumaticity in Triassic archosaurs, and the early evolution of the avian respiratory system. *PLoS One* **7**: e34094.

Byers, G. W. (1969). Evolution of wing reduction in crane flies (Diptera—Tipulidae). *Evolution* **23**: 346–354.

Byrnes, G. and A. J. Spence (2011). Ecological and biomechanical insights into the evolution of gliding in mammals. *Integrative and Comparative Biology* **51**: 991–1001.

Campbell, K. E. and E. P. Tonni (1983). Size and locomotion in teratorns (Aves, Teratornithidae). *Auk* **100**: 390–403.

Caple, G. R., R. T. Balda and W. R. Willis (1983). The physics of leaping animals and the evolution of pre-flight. *American Naturalist* **121**: 455–467.

Carpenter, G. H. (1913). *The Life-Story of Insects*. G. P. Putnam's Sons, New York. 134 pp.

Carpenter, F. M. (1992). Superclass Hexapoda. In *Treatise on Invertebrate Paleontology, Part R, Arthopoda 3–4*, R. L. Kaesler, Ed. Geological Society of America, Boulder, Colorado. 655 pp.

Chapman, R. F. (1982). *The Insects: Structure and Function*. Harvard University Press, Cambridge, Massachusetts. 919 pp.

Chatterjee, S. and R. J. Templin (2004). Posture, locomotion and paleoecology of pterosaurs. *Geological Society of America Special Papers* **376**: 1–64.

Chen, P. J., Z. M. Dong and S. N. Zhen (1998). An exceptionally well-preserved theropod dinosaur from the Yixian Formation of China. *Nature* **391**: 147–152.

Chiappe, L. M. and C. A. Walker (2002). Skeletal morphology and systematics of the Cretaceous Euenantiornithes (Ornithothoraces: Enantiornithese). In *Mesozoic Birds: Above the Heads of Dinosaurs*, L. M. Chiappe and L. M. Witmer, Eds. University of California Press, Berkeley, California. pp. 240–267.

Christiansen, P. and N. Bonde (2000). Axial and appendicular pneumaticity in *Archaeopteryx*. *Proceedings of the Royal Society of London Series B-Biological Sciences* **267**: 2501–2505.

Clarke, A. and H. O. Pörtner (2010). Temperature, metabolic power and the evolution of endothermy. *Biological Reviews* **85**: 703–727.

Colbert, E. H. (1967). Adaptations for gliding in the lizard *Draco*. *American Museum Novitates* **No. 2283**: 1–20.

Colbert, E. H. (1970). The Triassic gliding reptile *Icarosaurus*. *Bulletin of the American Museum of Natural History* **143**: 89–142.

Collini, C. A. (1784). Sur quelques zoolithes du Cabinet d'Histoire Naturelle de S.A.S.E. Palatine et de Baviére, à Mannheim. *Acta Academiae Theodoro-Palatinae, Mannheim, Pars Physica* **5**: 58–103.

Craik, S. R., J.-P. L. Savard and R. D. Titman (2009). Wing and body molts of male Red-breasted Mergansers in the Gulf of St. Lawrence, Canada. *Condor* **111**: 71–80.

Currie, P. J. (1991). *The Flying Dinosaurs*. Red Deer College Press, Red Deer, Alberta, Canada. 160 pp.

Cuvier, G. (1801). Extrait d'un ouvrage sur les espèces de quadrupèdes dont on a trouvé les ossmens dans l'intérieur de la terre. *Journal de Physique, de Chimie et d'Histoire Naturelle* **52**: 253–267.

Cuvier, G. (1809). Mémoire sur le squelette fossile d'un reptile volant des environs d'Aichstedt, que quelques naturalistes ont pris pour un oiseau, et dont nous formons un genre de Sauriens, sous le nom de Ptero-Dactyle. *Annales du Muséum national d'Histoire naturelle, Paris* **13**: 424–437.

Darwin, C. (1860). *On the Origin of Species*. J. Murray, London. 502 pp.

Davenport, J. (1994). How and why do flying fish fly? *Reviews in Fish Biology and Fisheries* **4**: 184–214.

Dial, K. P. (2003). Wing-assisted incline running and the evolution of flight. *Science* **299**: 402–404.

Dial, K. P., B. E. Jackson and P. Segre (2008). A fundamental avian wing-stroke provides a new perspective on the evolution of flight. *Nature* **451**: 985–989.

Dial, K. P., R. J. Randall and T. R. Dial (2006). What use is half a wing in the ecology and evolution of birds? *BioScience* **56**: 437–445.

Douglas, M. M. (1981). Thermoregulatory significance of thoracic lobes in the evolution of insect wings. *Science* **211**: 84–86.

Dudley, R. (2000). *The Biomechanics of Insect Flight: Form, Function, Evolution*. Princeton University Press, Princeton, New Jersey. 476 pp.

Dudley, R., G. Byrnes, S. P. Yanoviak, B. Borrell, R. M. Brown and J. A. McGuire (2007). Gliding and the functional origins of flight: Biomechanical novelty or necessity? In *Annual Review of Ecology, Evolution, and Systematics*. Annual Reviews, Palo Alto. vol. 38, pp. 179–201.

Dunnigan, J. F. (2003). *How to Make War: A Comprehensive Guide to Modern Warfare in the Twenty-first Century*. HarperCollins, New York. 672 pp.

Dyke, G. J., R. L. Nudds and J. M. V. Rayner (2006). Flight of *Sharovipteryx mirabilis*: The world's first delta-winged glider. *Journal of Evolutionary Biology* **19**: 1040–1043.

Edwards, J. S. (1986). Arthropods as pioneers: Recolonization of the blast zone on Mt. St. Helens. *Northwest Environmental Journal* **2**: 63–73.

Elgin, R. A., D. W. E. Hone and E. Frey (2011). The extent of the pterosaur flight membrane. *Acta Palaeontologica Polonica* **56**: 99–111.

Engel, M. S., S. R. Davis and J. Prokop (2013). Insect wings: The evolutionary developmental origins of Nature's first flyers. In *Arthropod Biology and Evolution: Molecules, Development, Morphology*. A. Minelli, G. Boxshall, and G. Fusco, Eds. Springer Verlag, Berlin. pp. 269–298.

Engel, M. S. and D. A. Grimaldi (2004). New light shed on the oldest insect. *Nature* **427**: 627–630.

Evans, J. (1865). One portion of a cranium and of a jaw in the slab containing the fossil remains of *Archaeopteryx*. *Natural History Review, series 2* **5**: 415–421.

Evenhuis, N. L. (1997). Review of flightless Dolichopodidae (Diptera) in the Hawaiian Islands. *Bishop Museum Occasional Papers* **53**: 1–30.

Farmer, C. G. and K. Sanders (2010). Unidirectional airflow in the lungs of alligators. *Science* **327**: 338–340.

Feduccia, A. (1996). *The Origin and Evolution of Birds*. Yale University Press, New Haven, Connecticut. 420 pp.

Feduccia, A., T. Lingham-Soliar and J. R. Hinchliffe (2005). Do feathered dinosaurs exist? Testing the hypothesis on neontological and paleontological evidence. *Journal of Morphology* **266**: 125–166.

Fenton, M. B., D. Audet, M. K. Obrist and J. Rydell (1995). Signal strength, timing, and self-deafening—The evolution of echolocation in bats. *Paleobiology* **21**: 229–242.

Fisher, H. I. (1957). Bony mechanism of automatic flexion and extension in the pigeon's wing. *Science* **126**: 446.

Flower, J. W. (1964). On the origin of flight in insects. *Journal of Insect Physiology* **10**: 81–88.

Forbes, W. T. M. (1943). The origin of wings and venational types in insects. *American Midland Naturalist* **29**: 381–405.

Forsman, K. A. and M. G. Malmquist (1988). Evidence for echolocation in the common shrew *Sorex araneus*. *Journal of Zoology* (London) **216**: 655–662.

Fox, R., S. W. Lehmkuhle and D. H. Westendorf (1976). Falcon visual acuity. *Science* **192**: 263–265.

Fraser, N. C., P. E. Olsen, A. C. Dooley Jr. and T. R. Ryan (2007). A new gliding tetrapod (Diapsida: ?Archosauromorpha) from the Upper Triassic (Carnian) of Virginia. *Journal of Vertebrate Paleontology* **27**: 261–265.

Frey, E., H.-D. Sues and W. Munk (1997). Gliding mechanism in the Late Permian reptile *Coelurosauravus*. *Science* **275**: 1450–1452.

Garner, J. P., G. K. Taylor and A. L. R. Thomas (1999). On the origins of birds: The sequence of character acquisition in the evolution of avian flight. *Proceedings of the Royal Society of London Series B-Biological Sciences* **266**: 1259–1266.

Garrouste, R., G. Clement, P. Nel, M. S. Engel, P. Grandcolas, C. D'Haese, L. Lagebro, J. Denayer, P. Gueriau, P. Lafaite, S. Olive, C. Prestianni and A. Nel (2012). A complete insect from the Late Devonian period. *Nature* **488**: 82–85.

Gauthier, J. A. (1986). Saurischian monophyly and the origin of birds. In *The Origin of Birds and the Evolution of Flight*, K. Padian, Ed. California Academy of Science, San Francisco. pp. 1–55.

Gegenbaur, C. (1870). *Grundzüge der vergleichenden Anatomie*. Wilhelm Engelmann, Leipzig. 890 pp.

Gegenbaur, C. (1878). *Elements of Comparative Anatomy*. Macmillan and Co., London. 645 pp.

Giannini, N. P. (2012). Toward an integrative theory on the origin of bat flight. In *Evolutionary History of Bats: Fossils, Molecules and Morphology*, G. F. Gunnell and N. B. Simmons, Eds. Cambridge University Press, Cambridge. pp. 353–384.

Giebel, C. (1877). Neueste Entdeckung einer zweiten *Archaeopteryx lithographica*. *Zeitschrift für die Gesammten Naturwissenschaften* **49**: 326–327.

Godefroit, P., A. Cau, H. Dong-Yu, F. Escuillie, W. Wenhao and G. Dyke (2013). A Jurassic avialan dinosaur from China resolves the early phylogenetic history of birds. *Nature* **498**: 359–362.

Gould, E., N. C. Negus and A. Novick (1964). Evidence for echolocation in shrews. *Journal of Experimental Zoology* **156**: 19–37.

Gould, S. J. (1985). Not necessarily a wing. *Natural History* **94**: 12–25.

Grimaldi, D. A. (2010). 400 million years on six legs: On the origin and early evolution of Hexapoda. *Arthropod Structure & Development* **39**: 191–203.

Grimaldi, D. A. and M. S. Engel (2005). *Evolution of the Insects*. Cambridge University Press, New York. 755 pp.

Gullan, P. J. and P. S. Cranston (2010). *The Insects: An Outline of Entomology*. Wiley-Blackwell, Hoboken, New Jersey. 565 pp.

Hasenfuss, I. (2002). A possible evolutionary pathway to insect flight starting from lepismatid organization. *Journal of Zoological Systematics and Evolutionary Research* **40**: 65–81.

Hedrick, T. L. (2011). Damping in flapping flight and its implications for manoeuvring, scaling and evolution. *Journal of Experimental Biology* **214**: 4073–4081.

Heilmann, G. (1927). *The Origin of Birds*. D. Appleton, New York. 209 pp.

Heinicke, M. P., E. Greenbaum, T. R. Jackman and A. M. Bauer (2012). Evolution of gliding in Southeast Asian geckos and other vertebrates is temporally congruent with dipterocarp forest development. *Biology Letters* **8**: 994–997.

Henderson, D. M. (2010). Pterosaur body mass estimates from three-dimensional mathematical slicing. *Journal of Vertebrate Paleontology* **30**: 768–785.

Hennig, W. (1981). *Insect Phylogeny*. John Wiley, New York. 514 pp.

Heptonstall, W. B. (1971). An analysis of the flight of the Cretaceous pterodactyl *Pteranodon ingens* (March). *Scottish Journal of Geology* **7**: 61–78.

Hone, D. W. E. (2012). Pterosaur research: Recent advances and a future revolution. *Acta Geologica Sinica-English Edition* **86**: 1366–1376.

Hone, D. W. E. and M. J. Benton (2007). An evaluation of the phylogenetic relationships of the pterosaurs among archosauromorph reptiles. *Journal of Systematic Palaeontology* **5**: 465–469.

Hone, D. W. E. and D. M. Henderson (2014). The posture of floating pterosaurs: Ecological implications for inhabiting marine and freshwater habitats. *Palaeogeography, Palaeoclimatology, Palaeoecology* **394**: 89–98.

Hornschemeyer, T., J. T. Haug, O. Béthoux, R. G. Beutel, S. Charbonnier, T. A. Hegna, M. Koch, J. Rust, S. Wedmann, S. Bradler and R. Willmann (2013). Is *Strudiella* a Devonian insect? *Nature* **494**: E3–E4.

Hu, D., L. Hou, L. Zhang and X. Xu (2009). A pre-*Archaeopteryx* troodontid theropod from China with long feathers on the metatarsus. *Nature* **461**: 640–643.

Huene, F.v. (1914). Beitrage zur Geschichte der Archosaurier. *Geologische und Paläontologische Abhandlungen N.F.* **13**: 1–53.

Hutcheon, J. M., J. A. W. Kirsch and J. D. Pettigrew (1998). Base-compositional biases and the bat problem. III. The question of microchiropteran monophyly. *Philosophical Transactions of the Royal Society of London Series B-Biological Sciences* **353**: 607–617.

Jepsen, G. L. (1966). Early Eocene bat from Wyoming. *Science* **154**: 1333–1339.

Jex, H. R. (2000). Making pterodactyls fly (QN Story). Published by TWITT (The Wing Is the Thing). Retrieved 5/12/2008, from http://www.twitt.org/QNStory.html.

Ji, Q., P. J. Currie, M. A. Norell and S.-A. Ji (1998). Two feathered dinosaurs from northeastern China. *Nature* **393**: 753–761.

Jones, K. J. and H. H. Genoways (1970). Chiropteran systematics. In *About Bats: A Chiropteran Symposium*, R. H. Slaughter and D. W. Walton, Eds. Southern Methodist University Press, Dallas. pp. 3–21.

Jusufi, A., Y. Zeng, R. J. Full and R. Dudley (2011). Aerial righting reflexes in flightless animals. *Integrative and Comparative Biology* **51**: 937–943.

King, A. S. and D. Z. King (1979). Avian morphology: General principles. In *Form and Function in Birds*, A. S. King and J. McLelland, Eds. Academic Press, New York. vol. 1, pp. 1–38.

Kingsolver, J. G. and M. A. R. Koehl (1985). Aerodynamics, thermoregulation, and the evolution of insect wings: Differential scaling and evolutionary change. *Evolution* **39**: 488–504.

Kingsolver, J. G. and M. A. R. Koehl (1994). Selective factors in the evolution of insect wings. *Annual Review of Entomology* **39**: 425–451.

Koopman, K. F. (1984). A synopsis of the families of bats. Part VII. *Bat Research News* **25**: 25–27.

Kukalova-Peck, J. (1978). Origin and evolution of insect wings and their relation to metamorphosis, as documented by fossil record. *Journal of Morphology* **156**: 53–125.

Kukalova-Peck, J. (1983). Origin of the insect wing and wing articulation from the arthropodan leg. *Canadian Journal of Zoology* **61**: 1618–1669.

Kukalova-Peck, J. (1985). Ephemeroid wing venation based upon new gigantic Carboniferous mayflies and basic morphology phylogeny and metamorphosis of pterygote insects (Insecta, Ephemerida). *Canadian Journal of Zoology* **63**: 933–955.

Kukalova-Peck, J. (1997). Arthropod phylogeny and "basal" morphological structures. In *Arthropod Relationships*, R. A. Fortey and R. H. Thomas, Eds. Chapman and Hall, London. pp. 269–279.

Lee, M. S. Y. and T. H. Worthy (2012). Likelihood reinstates *Archaeopteryx* as a primitive bird. *Biology Letters* **8**: 299–303.

Li, P.-P., K.-Q. Gao, L.-H. Hou and X. Xu (2007). A gliding lizard from the Early Cretaceous of China. *Proceedings of the National Academy of Sciences of the United States of America* **104**: 5507–5509.

Lingham-Soliar, T. (2003). The dinosaurian origin of feathers: Perspectives from dolphin (Cetacea) collagen fibers. *Naturwissenschaften* **90**: 563–567.

Lingham-Soliar, T. (2010). Dinosaur protofeathers: Pushing back the origin of feathers into the Middle Triassic? *Journal of Ornithology* **151**: 193–200.

Lü, J. C., D. M. Unwin, D. C. Deeming, X. S. Jin, Y. Q. Liu and Q. A. Ji (2011). An egg-adult association, gender, and reproduction in pterosaurs. *Science* **331**: 321–324.

Maina, J. N. (2000). What it takes to fly: The structural and functional respiratory refinements in birds and bats. *Journal of Experimental Biology* **203**: 3045–3064.

Maiorana, V. C. (1979). Why do adult insects not moult? *Biological Journal of the Linnean Society* **11**: 253–258.

Makovicky, P. J. and P. J. Currie (1998). The presence of a furcula in tyrannosaurid theropods, and its phylogenetic and functional implications. *Journal of Vertebrate Paleontology* **18**: 143–149.

Marden, J. H. (2003). The surface-skimming hypothesis for the evolution of insect flight. *Acta Zoologica Cracoviensia* **46**: 73–84.

Marden, J. H. and M. G. Kramer (1994). Surface-skimming stoneflies: A possible intermediate stage in insect flight evolution. *Science* **266**: 427–430.

Marden, J. H. and M. G. Kramer (1995). Locomotor performance of insects with rudimentary wings. *Nature* **377**: 332–334.

Marden, J. H., M. R. Wolf and K. E. Weber (1997). Aerial performance of Drosophila melanogaster from populations selected for upwind flight ability. *Journal of Experimental Biology* **200**: 2747–2755.

Martin, L. D. (1983). The origin and early radiation of birds. In *Perspectives in Ornithology*, A. H. Brush and G. A. Clark Jr., Eds. Cambridge University Press, London. pp. 291–353.

Martin, L. D., J. D. Stewart and K. N. Whetstone (1980). The origin of birds: Structure of the tarsus and teeth. *Auk* **97**: 86–93.

Maynard Smith, J. (1952). The importance of the nervous system in the evolution of animal flight. *Evolution* **6**: 127–129.

McGuire, J. A. and R. Dudley (2011). The biology of gliding in flying lizards (genus *Draco*) and their fossil and extant analogs. *Integrative and Comparative Biology* **51**: 983–990.

McMahon, T. A. (1984). *Muscles, Reflexes, and Locomotion*. Princeton University Press, Princeton, New Jersey. 331 pp.

McNab, B. K. (1994). Energy conservation and the evolution of flightlessness in birds. *American Naturalist* **144**: 628–642.

Medeiros, M. J. (2008). A new species of flightless, jumping, alpine moth of the genus Thyrocopa from Hawaii (Lepidoptera: Xyloryctidae: Xyloryctinae). *Zootaxa* **1830**: 57–62.

Meng, J., Y. Hu, Y. Wang, X. Wang and C. Li (2006). A mesozoic gliding mammal from northeastern China. *Nature* **444**: 889–893.

Müller, F. (1873). Beiträge zur Kenntniss der Termiten. *Jenaische Zeitschrift für Naturwissenschaft* **7**: 333–358; 451–463.

Müller, F. (1875). Beiträge zur Kenntniss der Termiten. *Jenaische Zeitschrift für Naturwissenschaft* **9**: 241–264.

Murphy, W. J., E. Eizirik, W. E. Johnson, Y. P. Zhang, O. A. Ryder and S. J. O'Brien (2001). Molecular phylogenetics and the origins of placental mammals. *Nature* **409**: 614–618.

Nesbitt, S. J. (2011). The early evolution of archosaurs: Relationships and the origin of major clades. *Bulletin of the American Museum of Natural History* **352**: 1–292.

Nettleship, D. N. and P. G. H. Evans (1985). Distribution and status of the Altlantic Alcidae. In *The Atlantic Alcidae: The Evolution, Distribution and Biology of the Auks Inhabiting the Atlantic Ocean and Adjacent Water Areas*, D. N. Nettleship and T. R. Birkhead, Eds. Academic Press, New York. pp. 53–154.

Neuweiler, G. (2000). *The Biology of Bats*. Oxford University Press, New York. 310 pp.

Newman, B. G., S. B. Savage and D. Schoulla (1977). Model tests on a wing section of an *Aeschna* dragonfly. In *Scale Effects in Animal Locomotion*, T. J. Pedley, Ed. Academic Press, New York. pp. 445–477.

Newton, I. (2008). *The Migration Ecology of Birds*. Academic Press, Boston. 976 pp.

Norberg, U. M. (1985). Evolution of flight in birds: Aerodynamic, mechanical and ecological aspects. In *The Beginnings of Birds*, M. K. Hecht, J. H. Ostrom, G. Viohl, and P. Wellnhofer, Eds. Freunde des Jura-Museums Eichstätt, Eichstätt, Germany. pp. 293–303.

Norberg, U. M. (1985). Evolution of vertebrate flight: An aerodynamic model for the transition from gliding to active flight. *American Naturalist* **126**: 303–327.

Norberg, U. M. (1990). *Vertebrate Flight: Mechanics, Physiology, Morphology, Ecology and Evolution*. Springer-Verlag, Berlin. 291 pp.

Novacek, M. J. (1985). Evidence for echolocation in the oldest known bats. *Nature* **315**: 140–141.

Novacek, M. J. (2001). Mammalian phylogeny: Genes and supertrees. *Current Biology* **11**: R573–R575.

Novacek, M. J. and A. R. Wyss (1986). Higher-level relationships of the recent eutherian orders: Morphological evidence. *Cladistics* **2**: 257–287.

Nudds, R. L. and G. J. Dyke (2009). Forelimb posture in dinosaurs and the evolution of the avian flapping flight-stroke. *Evolution* **63**: 994–1002.

O'Connor, P. M. (2009). Evolution of archosaurian body plans: Skeletal adaptations of an air-sac-based breathing apparatus in birds and other archosaurs. *Journal of Experimental Zoology Part a-Ecological Genetics and Physiology* **311A**: 629–646.

Oliver, J. A. (1951). "Gliding" in amphibians and reptiles, with a remark on an arboreal adaptation in the lizard, *Anolis carolinensis* Voigt. *American Naturalist* **85**: 171–176.

Ostrom, J. H. (1975). The origin of birds. *Annual Review of Earth and Planetary Sciences* **3**: 55–77.

Ostrom, J. H. (1976). *Archaeopteryx* and the origin of birds. *Biological Journal of the Linnean Society* **8**: 91–182.

Padian, K. (1982). Running, leaping, lifting off—flight evolved from the ground up, not the trees down. *Sciences-New York* **22**: 10–15.

Padian, K. (1983). A functional analysis of flying and walking in pterosaurs. *Paleobiology* **9**: 218–239.

Padian, K. (1984). The origin of pterosaurs. In *Third Symposium on Mesozoic Terrestrial Ecosystems*, W. E. Reif and F. Westfall, Eds. Attempto Verlag, Tübingen. pp. 163–168.

Padian, K. (1985). The origins and aerodynamics of flight in extinct vertebrates. *Palaeontology* **28**: 413–434.

Padian, K. (1991). Pterosaurs: Were they functional birds or functional bats. In *Biomechanics in Evolution*, J. M. V. Rayner and R. J. Wootton, Eds. Cambridge University Press, Cambridge. pp. 146–160.

Padian, K. and J. M. V. Rayner (1993). The wings of pterosaurs. *American Journal of Science* **293A**: 91–166.

Paul, G. S. (2002). *Dinosaurs of the Air: The Evolution and Loss of Flight in Dinosaurs and Birds*. Johns Hopkins University Press, Baltimore. 460 pp.

Pennycuick, C. J. (1972). *Animal Flight*. Edward Arnold, London. 68 pp.

Pennycuick, C. J., T. Alerstam and B. Larsson (1979). Soaring migration of the common crane *Grus grus* observed by radar and from an aircraft. *Ornis Scandinavica* **10**: 241–251.

Peterson, K., P. Birkmeyer, R. Dudley and R. S. Fearing (2011). A wing-assisted running robot and implications for avian flight evolution. *Bioinspiration & Biomimetics* **6**: 046008.

Pettigrew, J. D. (1986). Flying primates? Megabats have the advanced pathway from eye to midbrain. *Science* **231**: 1304–1306.

Pettigrew, J. D. (1991). A fruitful, wrong hypothesis—Response. *Systematic Zoology* **40**: 231–239.

Pettigrew, J. D. (1991). Wings or brain—convergent evolution in the origins of bats. *Systematic Zoology* **40**: 199–216.

Pettigrew, J. D. (1995). Flying primates: Crashed, or crashed through? In *Ecology, Evolution and Behaviour of Bats*, P. A. Racey and S. M. Swift, Eds. pp. 3–26.

Pettigrew, J. D., B. G. M. Jamieson, S. K. Robson, L. S. Hall, K. I. McAnally and H. M. Cooper (1989). Phylogenetic relations between microbats, megabats and primates (Mammalia: Chiroptera and Primates). *Philosophical Transactions of the Royal Society of London Series B-Biological Sciences* **325**: 489–559.

Powers, L. V., S. C. Kandarian and T. H. Kunz (1991). Ontogeny of flight in the little brown bat, *Myotis lucifugus*— behavior, morphology, and muscle histochemistry. *Journal of Comparative Physiology a-Sensory Neural and Behavioral Physiology* **168**: 675–685.

Rayner, J. M. V. (1985a). Mechanical and ecological constraints on flight evolution. In *The Beginnings of Birds*, M. K. Hecht, J. H. Ostrom, G. Viohl and P. Wellnhofer, Eds. Freunde des Jura-Museums Eichstätt, Eichstätt, Germany. pp. 279–288.

Rayner, J. M. V. (1985b). Cursorial gliding in protobirds: An expanded version of a discussion contribution. In *The Beginnings of Birds*, M. K. Hecht, J. H. Ostrom, G. Viohl, and P. Wellnhofer, Eds. Freunde des Jura-Museums Eichstätt, Eichstätt, Germany. pp. 289–292.

Rayner, J. M. V. (1988). The evolution of vertebrate flight. *Biological Journal of the Linnean Society* **34**: 269–287.

Rayner, J. M. V. (1988). Form and function of avian flight. *Current Ornithology* **5**: 1–66.

Rayner, J. M. V. (1989). Vertebrate flight and the origins of flying vertebrates. In *Evolution and the Fossil Record*, K. Allen and D. Briggs, Eds. Belhaven Press, London. pp. 188–217.

Rayner, J. M. V. (1999). Estimating power curves of flying vertebrates. *Journal of Experimental Biology* **202**: 3449–3461.

Rees, C. J. C. (1975). Aerodynamic properties of an insect wing section and a smooth aerofoil compared. *Nature* **258**: 141–142.

Renesto, S. and G. Binelli (2006). *Vallesaurus cenensis* Wild, 1991, a drepanosaurid (Reptilia, Diapsida) from the late Triassic of northern Italy. *Rivista Italiana Di Paleontologia E Stratigrafia* **112**: 77–94.

Robinson, P. L. (1967). Triassic vertebrates from lowland and upland. *Science & Culture* **33**: 169–173.

Roff, D. A. (1984). The cost of being able to fly—a study of wing polymorphism in 2 species of crickets. *Oecologia* **63**: 30–37.

Roff, D. A. (1990). The evolution of flightlessness in insects. *Ecological Monographs* **60**: 389–421.

Ross, A. J. (2010). A review of the Carboniferous fossil insects from Scotland. *Scottish Journal of Geology* **46**: 157–168.

Ruben, J. A., T. D. Jones and N. R. Geist (2003). Respiratory and reproductive paleophysiology of dinosaurs and early birds. *Physiological and Biochemical Zoology* **76**: 141–164.

Russell, A. P., L. D. Dijkstra and G. L. Powell (2001). Structural characteristics of the patagium of *Ptychozoon kuhli* (Reptilia: Gekkonidae) in relation to parachuting locomotion. *Journal of Morphology* **247**: 252–263.

Schmidt-Nielsen, K. (1972). Locomotion: Energy cost of swimming, flying and running. *Science* **177**: 222–227.

Schmidt-Nielsen, K. (1990). *Animal Physiology: Adaptation and Environment*. Cambridge University Press, Cambridge. 602 pp.

Seebacher, F. (2003). Dinosaur body temperatures: The occurrence of endothermy and ectothermy. *Paleobiology* **29**: 105–122.

Seeley, H. G. (1901). *Dragons of the Air, an Account of Extinct Flying Reptiles*. Methuen & Co., London. 239 pp.

Shear, W. A., P. M. Bonamo, J. D. Grierson, W. D. I. Rolfe, E. L. Smith and R. A. Norton (1984). Early land animals in North America: Evidence from Devonian age arthropods from Gilboa, New York. *Science* **224**: 492–494.

Shipman, P. (1998). *Taking Wing: Archaeopteryx and the Evolution of Bird Flight*. Simon & Schuster, New York. 336 pp.

Shoshani, J. and M. C. McKenna (1998). Higher taxonomic relationships among extant mammals based on morphology, with selected comparisons of results from molecular data. *Molecular Phylogenetics and Evolution* **9**: 572–584.

Simmons, N. B. (1995). Bat relationships and the origin of flight. *Symposium of the Zoological Society of London* **67**: 27–43.

Simmons, N. B. and J. H. Geisler (1998). Phylogenetic relationships of *Icaronycteris*, *Archaeonycteris*, *Hassianycteris*, and *Palaeochiropteryx* to extant bat lineages, with comments on the evolution of echolocation and foraging strategies in Microchiroptera. *Bulletin of the American Museum of Natural History* **235**: 1–182.

Simmons, N. B., M. J. Novacek and R. J. Baker (1991). Approaches, methods, and the future of the chiropteran monophyly controversy—Reply. *Systematic Zoology* **40**: 239–243.

Simmons, N. B. and T. H. Quinn (1994). Evolution of the digital tendon locking mechanism in bats and dermopterans: A phylogenetic perspective. *Journal of Mammalian Evolution* **2**: 231–254.

Simmons, N. B., K. L. Seymour, J. Habersetzer and G. F. Gunnell (2008). Primitive Early Eocene bat from Wyoming and the evolution of flight and echolocation. *Nature* **451**: 818–821.

Slikas, B., S. L. Olson and R. C. Fleischer (2002). Rapid, independent evolution of flightlessness in four species of Pacific Island rails (Rallidae): An analysis based on mitochondrial sequence data. *Journal of Avian Biology* **33**: 5–14.

Smith, J. D. (1977). Comments on flight and the evolution of bats. In *Major Patterns in Vertebrate Evolution*, M. K. Hecht, P. C. Goody, and B. M. Hecht, Eds. Plenum Press, New York. pp. 427–437.

Socha, J. J. and M. LaBarbera (2005). Effects of size and behavior on aerial performance of two species of flying snakes (*Chrysopelea*). *Journal of Experimental Biology* **208**: 1835–1847.

Soemmerring, S. T. v. (1817). Über einen *Ornithocephalus brevirostris* der Vorwelt. *Denkschriften der koniglichen bayerischen Akademie der Wissenschaften München, mathematisch-physikalische Classe* **6**: 89–104.

Song, A., X. D. Tian, E. Israeli, R. Galvao, K. Bishop, S. Swartz and K. Breuer (2008). Aeromechanics of membrane wings with implications for animal flight. *AIAA Journal* **46**: 2096–2106.

Speakman, J. R. (1993). Flight capabilities in *Archaeopteryx*. *Evolution* **47**: 336–340.

Speakman, J. R. and P. A. Racey (1991). No cost of echolocation for bats in flight. *Nature* **350**: 421–423.

Springer, M. S., M. J. Stanhope, O. Madsen and W. W. de Jong (2004). Molecules consolidate the placental mammal tree. *Trends in Ecology & Evolution* **19**: 430–438.

Springer, M. S., E. C. Teeling, O. Madsen, M. J. Stanhope and W. W. de Jong (2001). Integrated fossil and molecular data reconstruct bat echolocation. *Proceedings of the National Academy of Sciences of the United States of America* **98**: 6241–6246.

Stafford, B. J. (1999). *Taxonomy and Ecological Morphology of the Flying Lemurs (Dermoptera, Cynocephalidae)*. Ph.D. Dissertation, City University of New York. 464 pp.

Staniczek, A. H., P. Sroka and G. Bechly (2014). Neither silverfish nor fowl: The enigmatic Carboniferous *Carbotriplura kukalovae* Kluge, 1996 (Insecta: Carbotriplurida) is the putative fossil sister group of winged insects (Insecta: Pterygota). *Systematic Entomology*: DOI: 10.1111/syen.12076.

Swan, L. W. (1970). Goose of the Himalayas. *Natural History* **70**: 68–75.

Swartz, S. M., M. S. Groves, H. D. Kim and W. R. Walsh (1996). Mechanical properties of bat wing membrane skin. *Journal of Zoology* **239**: 357–378.

Tauber, C. A., M. J. Tauber and J. G. Giffin (2007). Flightless Hawaiian Hemerobiidae (Neuroptera): Comparative morphology and biology of a brachypterous species, its macropterous relative and intermediate forms. *European Journal of Entomology* **104**: 787–800.

Teeling, E. C. (2009). Hear, hear: The convergent evolution of echolocation in bats? *Trends in Ecology & Evolution* **24**: 351–354.

Teeling, E. C., O. Madsen, R. A. Van Den Bussche, W. W. de Jong, M. J. Stanhope and M. S. Springer (2002). Microbat paraphyly and the convergent evolution of a key innovation in Old World rhinolophoid microbats. *Proceedings of the National Academy of Sciences of the United States of America* **99**: 1431–1436.

Teeling, E. C., M. Scally, D. J. Kao, M. L. Romagnoli, M. S. Springer and M. J. Stanhope (2000). Molecular evidence regarding the origin of echolocation and flight in bats. *Nature* **403**: 188–192.

Tennekes, H. (1996). *The Simple Science of Flight: From Insects to Jumbo Jets*. MIT Press, Cambridge, Massachusetts. 137 pp.

Thewisen, J. G. M. and S. K. Babcock (1991). Distinctive cranial and cervical innervation of wing muscles: New evidence for bat monophyly. *Science* **251**: 934–936.

Thomas, S. P. (1987). The physiology of bat flight. In *Recent Advances in the Study of Bats*, M. B. Fenton, P. A. Racey, and J. M. V. Rayner, Eds. Cambridge University Press, New York. pp. 75–99.

Thomassen, H. A., S. Gea, S. Maas, R. G. Bout, J. J. J. Dirckx, W. F. Decraemer and G. D. E. Povel (2007). Do Swiftlets have an ear for echolocation? The functional morphology of Swiftlets' middle ears. *Hearing Research* **225**: 25–37.

Thorington, R. W. Jr. and L. R. Heaney (1981). Body proportions and gliding adaptations of flying squirrels (Petauristinae). *Journal of Mammalogy* **62**: 101–114.

Thulborn, R. A. (1973). Thermoregulation in dinosaurs. *Nature* **245**: 51–52.

Tobalske, B. W. and K. P. Dial (2007). Aerodynamics of wing-assisted incline running in birds. *Journal of Experimental Biology* **210**: 1742–1751.

Trueman, J. W. H., B. E. Pfeil, S. A. Kelchner and D. K. Yeates (2004). Did stick insects really regain their wings? *Systematic Entomology* **29**: 138–139.

Tucker, V. A. (1968). Respiratory physiology of house sparrows in relation to high-altitude flight. *Journal of Experimental Biology* **48**: 55–66.

Tucker, V. A., T. J. Cade and A. E. Tucker (1998). Diving speeds and angles of a gyrfalcon (*Falco rusticolus*). *Journal of Experimental Biology* **201**: 2061–2070.

Unwin, D. M. (1987). Pterosaur locomotion: Joggers or waddlers? *Nature* **327**: 13–14.

Unwin, D. M. (1997). Pterosaur tracks and the terrestrial ability of pterosaurs. *Lethaia* **29**: 373–386.

Unwin, D. M. (1999). Pterosaurs: Back to the traditional model? *Trends in Ecology & Evolution* **14**: 263–268.

Unwin, D. M. and N. N. Bakhurina (1994). *Sordes pilosus* and the nature of the pterosaur flight apparatus. *Nature* **371**: 62–64.

Unwin, D. M. and D. C. Deeming (2008). Pterosaur eggshell structure and its implications for pterosaur reproductive biology. *Zitteliana Reihe B* **28**: 199–207.

Vandenberghe, N., J. Zhang and S. Childress (2004). Symmetry breaking leads to forward flapping flight. *Journal of Fluid Mechanics* **506**: 147–155.

Vaughan, T. A., J. M. Ryan and N. J. Czaplewski (2000). *Mammalogy*. Saunders College Publishing, Philadelphia. 565 pp.

Videler, J. J. (2005). *Avian Flight*. Oxford University Press, New York. 269 pp.

Vogel, S. (1994). *Life in Moving Fluids: The Physical Biology of Flow*. Princeton University Press, Princeton, New Jersey. 467 pp.

Vogel, S. (2001). *Prime Mover: A Natural History of Muscle*. W.W. Norton, New York. 370 pp.

Walker, A. D. (1972). New light on origin of birds and crocodiles. *Nature* **237**: 257–263.

Warrick, D. R. and K. P. Dial (1998). Kinematic, aerodynamic and anatomical mechanisms in the slow, maneuvering flight of pigeons. *Journal of Experimental Biology* **201**: 655–672.

Wellnhofer, P. (1988). Terrestrial locomotion in pterosaurs. *Historical Biology* **1**: 3–16.

Wellnhofer, P. (1996). *The Illustrated Encyclopedia of Prehistoric Flying Reptiles: Pterosaurs*. Barnes & Noble Books, New York. 192 pp.

Whalley, P. and E. A. Jarzembowski (1981). A new assessment of *Rhyniella*, the earliest known insect, from the Devonian of Rhynie, Scotland. *Nature* **291**: 317.

Whiting, M. F., S. Bradler and T. Maxwell (2003). Loss and recovery of wings in stick insects. *Nature* **421**: 264–267.

Wiest, F. C. (1995). The specialized locomotory apparatus of the freshwater hatchetfish family Gasteropelecidae. *Journal of Zoology* **236**: 571–592.

Wigglesworth, V. B. (1963). Origin of wings in insects. *Nature* **197**: 97–98.

Wigglesworth, V. B. (1973). Evolution of insect wings and flight. *Nature* **246**: 127–129.

Wigglesworth, V. B. (1976). The evolution of insect flight. In *Insect Flight*, R. C. Rainey, Ed. Blackwell Scientific Publications, Oxford. pp. 255–269.

Wild, R. (1984). Pterosaurs from the upper Triassic of Italy. *Naturwissenschaften* **71**: 1–11.

Wiley, E. O. and B. S. Lieberman (2011). *Phylogenetics: The Theory of Phylogenetic Systematics*. Wiley-Blackwell, Hoboken, New Jersey. 406 pp.

Witton, M. P. and M. B. Habib (2010). On the size and flight diversity of giant pterosaurs, the use of birds as pterosaur analogues and comments on pterosaur flightlessness. *PLoS One* **5**: e13982.

Wootton, R. J. (1972). Nymphs of Palaeodictyoptera (Insecta) from the Westphalian of England. *Palaeontology* **15**: 662–675.

Wootton, R. J. (1986). An approach to the mechanics of pleating in dragonfly wings. *Journal of Experimental Biology* **125**: 361–372.

Wootton, R. J. and C. P. Ellington (1991). Biomechanics and the origin of insect flight. In *Biomechanics in Evolution*, J. M. V. Rayner and R. J. Wootton, Eds. Cambridge University Press, Cambridge. pp. 99–112.

Yanoviak, S. P. (2010). Gliding ants of the tropical forest canopy. In *Ant Ecology*, L. Lach, C. Parr, and K. Abbott, Eds. Oxford University Press, Oxford. pp. 223–224.

Yanoviak, S. P. and R. Dudley (2006). The role of visual cues in directed aerial descent of *Cephalotes atratus* workers (Hymenoptera: Formicidae). *Journal of Experimental Biology* **209**: 1777–1783.

Yanoviak, S. P., R. Dudley and M. Kaspari (2005). Directed aerial descent in canopy ants. *Nature* **433**: 624–626.

Yanoviak, S. P., M. Kaspari and R. Dudley (2009). Gliding hexapods and the origins of insect aerial behaviour. *Biology Letters* **5**: 510–512.

Yanoviak, S. P., Y. Munk and R. Dudley (2011). Evolution and ecology of directed aerial descent in arboreal ants. *Integrative and Comparative Biology* **51**: 944–956.

Yanoviak, S. P., Y. Munk, M. Kaspari and R. Dudley (2010). Aerial manoeuverability in wingless gliding ants (*Cephalotes atratus*). *Proceedings of the Royal Society of London Series B-Biological Sciences* **277**: 2199–2204.

Zheng, X. T., Z. H. Zhou, X. L. Wang, F. C. Zhang, X. M. Zhang, Y. Wang, G. J. Wei, S. Wang and X. Xu (2013). Hind wings in basal birds and the evolution of leg feathers. *Science* **339**: 1309–1312.

# INDEX

Note: Page numbers followed by "n" indicate footnotes. Page numbers in *italics* indicate illustrations, and may also include text discussion.

aerobic muscle, 24
aerobic scope, 121, 145
air combat, 13
air sacs, respiratory, 123
airfoil, 42, *43*
airspeed. *See* speed
albatross, 5, 8, *9*, 58
altitudes, bird migration, 6
anaerobic muscle, 24
angle of attack, 39, 40
antagonistic muscles. *See* muscles
ants, directed aerial descent in, 48–49, 64, 87
aphids, 99, 165, *166*
arboreal animals, directed aerial descent by, 87, 88
arboreal theory, 108, 110
    *Archaeopteryx* in, 109
    bats and, 135
    biomechanics supports, 111
    climbing as exaptation for, 174
    dinosaurs and, 120
    directed aerial descent supports, 174
    *Microraptor* supports, 115
    pouncing proavis, compared with, 118
    pterosaurs and, 154, 155
Archaeognatha, fossil, 75
*Archaeopteryx*, 103–105, *104*, 109
    and Avialae, *115*, 116
    *Dienonychus* and, 110
    evolutionary relationships of, *107*
    feathers of, *114*
    flight of, 27, 105, 109, 175
    on phylogenies, 106, 112
    on phylogeny, by Godefroit team, *115*
    skeletal specializations of, 126
    teeth of, 106
    thecodonts and, 105
Archonta, 131, *132*
Archosauria, 106, 147, 151, *152*, *153*
Arctic tern, migration distance, 7, 8
area, wing, 44. *See also* surface area
*Argentavis*, 9
*Arthropleura*, 94
*Arthurdactylus conandoylensis*, 12
articulation, wing, 30. *See also* shoulder joint; wing articulation
aspect ratio, 45, 48, 62, 63
auks, 13, 168
*Aurornis*, *115*, 116
automatic wing extension, 31, 33
Aves, 106. *See also* birds
Avialae, 106, *107*, *115*, 116

bat wings, *32–33*, 178
bats
    ancestry of, 180
    arboreal flight evolution in, 174
    in Archonta, 131
    automatic wing extension in, 33
    colugos and, 131, 135
    convergent evolution in, 139, 143
    echolocation by, 133, 134, 141, 142, 144

bats (*continued*)
    families, molecular phylogeny of, 140
    flight evolution in, 135–136, 143
    flight muscles power echolocation, 144
    flight origin in, 137–140, 173
    fossil skeletons of, 132
    front limb modifications of, *136*, 145
    growth and flight in, 157
    heart and lungs of, 145
    *Icaronycteris*, 132
    limited structural variation in, 176
    megabats and microbats, 130, 137, 140
    megabats and primates, 137
    microbats, not a single lineage in, 140
    migration distance of, 7
    *Onychonycteris*, 133
    on phylogeny, *132*, *133*, 137
    physiological modification to increase power of, 145
    poor fossil record of, 131, 135
    predation by, 14, 15, 142
    sensing (other than echolocation) in, 134
    shrews, relationship to, 132
    vision in, 175
    wing structure of, 32–33, 178
    Yangochiroptera, 141
    Yangochiroptera and Yinpterochiroptera, *138*, *140*
    Yinochiroptera, 141
    Yinpterochiroptera, 141
beak, birds', 125
bedbugs, secondarily flightless, 100
bee flies, 14, 99
bees, 15, 99
beetles, 1, *8*, 19, 99
Bennett, S. Christopher, 152, 158, 162
"Big Four" powered flyers, 4–5, *37*, 173
    pterosaurs, only extinct members, 147
biomechanics, 111, 118–119
biramous limbs (crustaceans), 82
bird
    aspect ratios, 62
    automatic wing extension, 31
    beak as adaptation, 125
    flight origin, 173
    hand and finger bones, 30
    largest, 9
    migration altitudes, 6
    pneumatic bones, 124
    shoulder joint, 123
    wing structure, 30–31, *31*
birds
    arboreal flight evolution of, 174
    *Archaeopteryx*, *104*, 106, 109, 110, 126. *See also Archaeopteryx*
    as archosaurs, 151
    Avialae, 106
    biomechanics and flight evolution theories for, 111
    *Confuciusornis*, 112, 126
    cormorants, Galapagos flightless, 167
    and crocodilians, 106, 123
    dinosaur clavicles and, 106
    as dinosaur subgroup, 110
    directed aerial descent in, 120
    endothermy in, 121
    feathered dinosaurs related to, 113
    feathers of, 19, 30, *114*, 117, 180. *See also* feathers
    flightless rails, 167
    flightlessness in, 167, 171
    growth and flight, 157
    largest, 9
    limited structural variation in, 176
    in Maniraptora, 106
    migration distances of, 7–8
    more primitive than *Archaeopteryx*, 116
    Neornithes, 127
    one-way lungs of, 122, 123, 180
    origin of, 5
    Ornithurae, 127
    pouncing proavis theory and, 118–119
    protowing of, 178
    skeletal specializations of, 125
    tails of, 125
    theropod-bird relationship (skepticism of), 110
    wing skeleton, 125
bones
    bat wing, *32*

bird wing, *31*
pneumatic, 124
pterosaur wing, 33, *34*
"bounded ignorance", 82–85, 86, 89
brain. *See* nervous system
bristletails, 75, 79

camber, 30, 42, 151, 156, 177
*Carbotriplurida*, 76, 89
*Caudipteryx*, 112, 113, 118, 119
China. *See* Liaoning fossil beds
Chiroptera, 137. *See also* bats
chord, wing, 45
*Chrysopelea*, 68
clavicles (dinosaur), 105, 106
*Coelurosauravus*, 64, *65*
cold-blooded animals. *See* ectotherms
collarbones. *See* clavicles
Collembola, 74, 75
colugos, 68
  and bats, 131, 135, *136*
  gliding, *69*
  on the ground, 71
  on phylogeny, *132*, *133*
  vision, similarities with megabats, 139
  wing loading and speed, 69
  wing structure, 68
complex structures, evolution of, 23
condors, 6, *9*, 58
*Confuciusornis*, 112, *115*, 126
control-configured flyers, 26
convergent evolution, 17, 139
  due to flight constraints, 176
  versus homology, 176
cormorants, Galpagos flightless, 167
cost of locomotion, flight versus walking, 3, 55, 71
courting in flight, 14
crocodilians, 106, 123, 151, *152*, *153*
Crustacea, limb terminology of, 82
cursorial theory, 109, 110
  *Archaeopteryx* and, 109
  biomechanics and, 111
  compared with pouncing proavis theory, 118
  flapping for thrust on ground, 111
  *Microraptor* and, 115
  pterosaurs and, 154
Cuvier, Georges, 148, 149

damselflies, 92, 93, 95
Darwin, Charles
  arboreal theory proposed by, 108
  bat flight evolution described by, 135
  evolution of complex structures, explained, 23
  loss of flight discussed by, 167
  use of tree diagram, 16
*Deinonychus*, 106, *107*, 108, 109
design in evolution, 28
Dial, Kenneth, 119
dinosaurs
  *Archaeopteryx* as transition from, 105
  as archosaurs, 151
  *Aurornis*, *115*, 116
  *Caudipteryx* and *Protarchaeopteryx*, 112, 113
  clavicles in, 105, 106
  *Deinonychus*, 106, *107*, 108
  endothermy in, 122
  feathered, 113, 117
  Liaoning fossil beds, 111, 112
  lungs of, one-way, 123, 180
  Maniraptora, 106, *107*
  *Microraptor*, 113, *114*, 115, 120
  phylogeny of, *107*, *152*, *153*
  pneumatic bones of, 124
  *Scleromochlus* and pterosaurs and, 151, *153*
  secondarily flightless, 116
  thermoregulation in, 117, 121, 122
  theropods, 106, *107*. *See also* theropods
  *Velociraptor*, 106
  very small, 120, 121
Diptera (true flies). *See* flies
directed aerial descent, 71, 87, 174
  discovery, 48–50
  as unifying feature, 174
  in vertebrates, 69, 120
distance
  long migrations, 7–8
  trades off against speed, 28
dodo, loss of flight, 167
downstroke, 51, 52, *53*
*Draco*, 60, *61*, 63, 67

drag, 29, 40
  induced, 44
  lift-to-drag ratio, 44
  shape and, 29
  and size, 45
  stall and, 41
  streamlining and, 29, 42, 46
  teminal velocity and, 71
  thrust overcoming, 50
  viscous versus pressure, 46
dragonfly, 92, 93, *95*, 171
  aerial predation by, 14
  courtship flight of, 14
  flapping pattern of, 95
  gliding of, 48
  laying eggs in flight, 14
  lift to drag ratio in, 45, 58
  Odonatoptera, 92–94
  paleopterous wing hinge of, 90
  wing structure of, 21
ducks, seasonal loss of flight, 165
Dudley, Robert, 49, 66, 87

eagle, 5, 6, 9, 58
echolocation, 141, 142
  evolution of, 143
  in fossil bats, 133
  multiple origins (microbats), 141
  and vision, 175
ectotherms, 121, 159
eggs, laying, in flight, 14
Ellington, Charles, 84
Enantiornithes, *107*, *115*, 127
endotherms, 117, 121–123, 159
energetics, flight, 3, 158
energy conservation, 55, 71
Engel, Michael, 75, 76
Ephemeroptera, 92. *See also* mayflies
evolution
  arthropod limb structure and, 82
  of bat front limb, *136*
  bat molecular phylogeny, *140*
  benefits of gliding, 70–71
  of complex structures, 23
  convergent, 139, 143, 162, 176
  design in, 28
  of echolocation, 143
  of endothermy in birds, 121
  experiments, 84, 85, 87
  of feathers, 113, 117

flight origin in insects, 76
gliding and, 47, 50, 55
historical constraints on, 22, 176
of insect tracheal system, 79
insect wing origin, theories, 77
of living animals, 178, 179
living animals as analogs for, 86, 119
of loss of flight ability, 100
of lungs, one-way, 123, 180
maneuverability and stability in, 27
megabat and microbat relationship, *140*
of Neornithes, 127
oribatid mites evolving new joint, 80
paranotal lobe theory in, 80–81
of pneumatic bones, 124
of pterosaur, no molecular phylogenies for, 161
rails, loss of flight, 167, 168, 179
relationships, 19
sexual selection, 81
evolutionary relationships, 16, 19. *See also* phylogenies (phylogenetic trees)
evolutionary trees, 16. *See also* phylogenies (phylogenetic trees)
exaptation, 22, 69, 88, 122, 174
exite or exopodite (crustacean limb), 81, 82
exoskeleton, insect, 35

falcon, 6, 11, 14
falling, 71, 174
feathered dinosaurs, 117
  *Aurornis*, 116
  *Caudipteryx* and *Protarchaeopteryx*, 112
  *Microraptor*, 113
  pouncing proavis theory, 118–119
feathers, 31
  *Archaeopteryx*, 104, *114*
  *Aurornis*, 116
  *Caudipteryx*, 112
  *Confuciusornis*, 126
  contour, 31
  on dinosaurs, 113, 122
  evolution of, 117
  as insulation, 117, 122
  *Microraptor*, asymmetrical, 113
  origin of, 180

pennaceous, 112
  on phylogenies, 19
  primary, 30, *31*, 113, *114*
  *Protarchaeopteryx*, 112
  symmetrical, *Caudipteryx*, 113
featherwing beetle, *8*
fish, 70
flapping
  and aerobic scope, 121
  in arboreal and cursorial theories, 108, 109
  *Archaeopteryx*, 109
  Avialae, defined by, 106
  bats, 137
  bats, on phylogeny, *140*
  Big Four, in common, 173
  bird lungs and power for, 123
  dragonfly pattern, 95
  Enantiornithes, 127
  in evolution of insect flight, 89
  forces, 24, 52, *53*
  forewings and hindwings held together, 99
  grasshopper pattern, 95
  helicopter (analogy), 50
  and hovering power, 54
  limits structural variation, 176
  muscles, 24, 144, 174
  similar for all flying animals, 24
  for thrust, 50–52
  for thrust on ground, 111, 120
  transition from walking, 174
  upstroke and downstroke movements, 51
  versus gliding, 55
  weak, 55, 72, 80, 89, 136, 156
  wing movements, *51*
  wing-assisted incline running and, 119
fleas, 100, 167, 169–171
flies, 15, 99
  flightless, 170
  halteres, 100
flight, 2
  benefits of, 3–4
  courting in, 14
  energy cost, 3
  evolutionary success and, 4
  flapping, Big Four, 4, 173
  gliding and evolution of, 72
  laying eggs in, 14
  maximum altitudes, 6
  maximum speed, 5, 6
  power requirements for, 24, 121
  powered, 2, 50
  sensing requirements of, 26, 61
  speed, 3, 5, 6, 54
  underwater, 13
  unpowered, 2
  vision, acute, requirement for, 26, 61, 72, 175
flight evolution
  arboreal theory in, 108, 111, 120, 135, 154, 174
  *Archaeopteryx* and competing theories, 109
  of bats, 135–143, *136*, *140*
  biomechanics and flight origin theories, 111
  bird wing skeleton modifications, 125
  bird-theropod relationship (skepticism of), 110
  confused with bird ancestry, 109
  cursorial theory, 108–110, 155
  directed aerial descent, 88, 120, 174
  feathers and, 117, 180
  flapping for thrust on ground, 111
  gill theory, 77–79
  gliding and, 47, 55, 72, 84, 88
  of gliding animals, 69, 72
  historical constraints and, 176
  insects, 76, 89
    models of ancestral, 82–85
    wing origin, traditonal theories, 77
    wings not from legs, 97
  Kukalova-Peck's, 81
  living animals as analogs for, 86, 119
  loss of flight ability, 100, 164
  modifications to nervous system for, 61, 175
  paranotal lobe theory, 80–81
  pouncing proavis theory, 118–119
  of pterosaurs, and arboreal theory, 154, 155
  of pterosaurs, and cursorial theory, 154, *155*
  of pterosaurs, birds and bats, potential interactions, 173

*Index* [201]

flight evolution (*continued*)
    requirement for acute vision in, 26, 61, 72, 175
    *Rhyniognatha* (ancient insect fossil), 75
    silverfish falling behavior and, 87
    skeletal specializations in birds for, 125
    solar collecting and insect protowings, 84
    surface-skimming theory, 86–87
    transition from walking to flapping, 174
    weak flapping and, 55, 72, 80, 89, 136, 156
    wing-assisted incline running theory, 119–120
flight muscles, soaring birds, 58
flightlessness, secondary. *See* loss of flight
flying animals, 8, 9, 10, 178
    amphibian, 13
    limited structural variation in, 175, 176
    maneuverability of, 27
    require good vision, 26
flying fish, 70
flying foxes, Old World. *See* megabats
flying lemurs, 68. *See also* colugos
flying lizards. *See* Draco
flying squirrels, 63, 68
force, flapping, 24, 52, 53. *See also* drag; lift; thrust; weight
fossils
    *Archaeopteryx*, 103–105, *104*
    bats, 131, 132
    bristletail, 75
    *Carbotriplurida*, 76
    *Caudipteryx* and *Protarchaeopteryx*, 112
    *Confuciusornis*, 112, 126
    earliest winged insects, 76
    Elmo fossil beds, 94
    Enantiornithes, 127
    evolutionary relationships and, 19
    feathered dinosaurs, 112
    giant insects, 94
    *Icaronycteris*, 132, 133
    incomplete record, 179
    insects and insect wings, 89
    Liaoning fossil beds, 111
    mayflies, 93
    mayfly nymphs, *98*
    *Microraptor*, 113, *114*, 115
    nymphs with large wing pads, *98*
    oldest springtails, 74
    Ornithurae, 127
    *Onychonycteris*, 133
    Palaeodictyopterida, *91*
    pterosaur, *150*, 158, 162
    pterosaur trackways, 155, 156
    *Rhyniognatha*, 75
    springtail, 75
frogs, gliding, 69
fruit bats, Old World. *See* megabats
fruit fly, 6, 58
furculas, 105, 106

Garner, Joseph, 118
gecko, gliding, 67
geese, 6, 7, 9
Gegenbaur, Carl, 77
genus (plural genera), 11
giant insects, Carboniferous, 94
gigantothermy, 122n
gill theory of insect wing origin, 77–79, 85, 98
    exites in, 81
    Kukalova-Peck's modification of, 81
    problems, 79
    Wigglesworth, Vincent, supporting, 81
gills, insect, 77, *78*
glide angle
    ants, 49, 64
    model, ancestral insect, 85
glide tests, insect models, 85, 87
gliders
    animals, 60
    in equilibrium glide, 62, *63*
    hypothetical ancestral insects, as models of, 84
    lift-to-drag ratio and glide angle, 47
gliding, 2, 47–50, *63*
    ants, glide angle of, 49
    arboreal and cursorial theories and, 108
    ballistic and aerodynamic phases, 62
    biomechanics of bird flight evolution and, 111

directed aerial descent and, 87, 88
*Draco*, 60
evolution of, 50
evolutionary benefits of, 70–71
and falls, 70, 71
favors large animals, 58
flight, as a form of, 72
in flight evolution, 174
    bats, 135, 136
    insects, 84, 88, 89
    pterosaurs, 154, 155
high lift-to-drag ratio improves, 48
*Microraptor*, 115
model tests, extinct gliding
    lizards, 66
in paranotal lobe theory, 80
powered by gravity, 47
small size and, 48
soaring, 48
versus flapping, 55
weight and speed in, 47
wing-assisted incline running
    omits, 119
gliding animals, 48, *61*
angles of attack in, high, 63
ants, 64
aspect ratios of, 63
bat ancestor as, 135
colugos, 68, *69*
don't soar, 62
*Draco*, 60
energy conservation and, 71
escape from predators by, 70
as evolutionary endpoint, 72
extinct lizards, *65*
extinct mammals, 66
extinct reptiles, 64, *65*, 67
flying fish, 70
flying squirrels, 68
frogs, 69
geckos, 67
glide paths of, 62, *63*
living today, 67–70
*Microraptor*, 115
nervous systems of, 61
non-equilibrium glides in, 62, *63*
snakes, 68
wing lengths constrained in, 63
wings of, 61
gliding lizards, extinct, *65*

gliding reptiles. *See* gliding animals
goose. *See* geese
grasshoppers, flapping pattern, 95
Great Auks, 13, 168, 169
great blue heron, 6
griffenflies ("giant dragonflies"), 93, 94
Grimaldi, David, 75
ground up theory. *See* cursorial theory
growth, pterosaur, effect on flight,
    157–158

halteres (fly stabilizers), 100
Hasenfuss, I., 87
hatchetfish, 70
hawks, 14, 15, 58
Heilmann, Gerhard, 105
historical constraints, 22, 176
homology, 176
homonomous wings, 95
honeybees, 6, 7, 11, 14
house flies. *See* flies
hovering, 12–13, 51
    power requirements for, 54
    size and, 12, 55, 56
    structural requirements for, 55
hummingbird, 52

*Icaronycteris*, 132, 133
*Icarosaurus*, 64, *65*
induced drag, 44
insect exoskeleton, 35
insect protowings, *83*
    as courtship displays, 80
    as exites of leg, 81
    as gills, 77
    as paranotal lobes, 80
    as solar collectors, 82
insect wings
    anatomy of, *36*
    articulation of, 36
    inability to molt, 93
    membrane of, 35
    not modified legs, 35
    pleating for stiffness, 35, *36*
    structure of, 35–37
    veins in, 35
insects
    age of flight origin of, 76, 173
    arboreal flight evolution in, 174
    bedbugs, 100

insects (*continued*)
  body form variation of, 176, 178
  Carboniferous giants, 94
  *Carbotriplurida* fossil, 76
  cold-loving, and loss of flight, 170
  earliest winged, 76
  Ephemeroptera, 92
  fleas, 100
  flight evolution in, 77, 79–81
  flight origin theories, traditional, 77
  fly halteres, 100
  forewings, dominant, 99
  fossils, 75, 76, 78, 89, *91*, 93, 94
  four-winged plan, 94, 99
  gill theory, 77–79, 81
  gills of, aquatic, 77, *78*
  gliding by, 48
  homonomous wings, 95
  immature, terminology, 78
  integrated flight evolution
      hypothesis, 89
  Kukalova-Peck's theory, 81
  largest known, 93, 94
  legs did not become wings in, 97, 178
  lice, 100
  locked forwings and hindwings, 99
  loss of flight in, 100, 167, 169, 171
  mayflies, 92
    fossils, 93
    immature, with gills, 78
    subimago, 92
  model tests of ancestral, 82–85
  nymphs and naiads, 78
  Odonatoptera, 92–94
  oldest fossils of, 74
  only fly as adults, 92, 157
  origin, 4, 74
  Palaeodictyopterida, *91*, 92
  paranotal lobe theory, 80–81
  phylogeny of major groups, *90*
  protowings as solar collectors, 84
  *Rhyniognatha* (ancient fossil), 75
  silverfish, 76. *See also* silverfish
  single functional wing pair in
      flies, 99
  size, versus vertebrates, 178
  species with flying and flightless
      individuals, 166
  stoneflies and surface skimming,
      86–87
  tracheal system of, 79
  uniramous limbs of, 82
  upstroke of, 52
  wing hinge
    neopterous, 90
    origin, 180
    paleopterous, 90
  wing origin, questions about, 179
insulation, 117, 122
interference, mechanical, wing
    hinges, 96
islands, loss of flight on, 167

June beetle, green, 1, 19

Kaspari, Michael, 87
Kingsolver, Joel, 82, 84
Koehl, Mimi, 82, 84
Koopman, Karl, 141
Kramer, Melissa, 86
*Kuehneosaurus*, 64, *65*
Kukalova-Peck, Jarmila, 81

L/D. *See* lift-to-drag ratio
lactic acid in muscles, 25
*Lagosuchus*, 152, *153*. *See also*
    *Marasuchus*
laminar flow, 46
largest flyers, *9*
larva (immature insect), 78
Laurasiatheria, on phylogeny, *133*, *140*
Liaoning fossil beds, 111, 112, 126
lice, loss of flight in, 100, 167, 169, 170
lift, 39, 40
  angle of attack and, 40
  Bernoulli's equation and, 40
  hovering and vertical takeoff, 12
  on insect protowings, 84, 89
  relative wind and, 40
  speed and, 43
  stall and, 41, *42*
  trailing edge orientation and, 41
lift-to-drag ratio, 44
  gliding, 47, 48
  maximum, 44n
  and size, 58
loss of flight
  bats and pterosaurs, absent in, 165
  in birds, 165, 168, 171
  energy conservation and, 168

in Great Auk, 168
in insects, 165, 166, 169, 170
on islands, 167
legs and, 165
ostriches and, 166
penguins and, 167, 168
reversal, improbability of, 170
seasonal, 165
social insect worker caste, 166
winged and wingless aphids, *166*
lungs (flight modifications of), 122, 145, 159, 180

mammals, 66, *132*, *133*
mandibles, dicondylic, 75
maneuverability, 27
in animal flight evolution, 27
bird tails and, 125
of birds, 44
of colugos, 69
of dragonflies, 96
predators, avoiding with, 175
pterosaurs, size and, 158
and stability, lack of built-in, 175
maneuvering, pterosaur head crests and, 160
Maniraptora, 106
*Deinonychus* and *Velociraptor*, 106
evolutionary relationships of, *107*
Liaoning fossil beds, 112
*Microraptor*, 113
on phylogeny, *107*, 115
secondarily flightless, 116
*Marasuchus*, 152n, *153*
Marden, James, 86
mating in flight, 14
mayflies, 92
feeding, 15
fossils, 93, *98*
gills on, immature, *78*
immature, *78*, *98*
mating in flight, 14
subimago, 92, 97
Maynard Smith, John, 27
McGuire, Jimmy, 66
*Mecistotrachelos*, 66, 67
megabats, 130, *138*
difference from microbats, 137

form a lineage with some microbats, 140
linked to primates, 137, 139
on molecular phylogeny, *140*
vision in, 134, 139
*Meganeuropsis*, 93
membrane, wing, 32, *34*, 35, *36*
microbats, 130, *138*
difference from megabats, 137
echolocation in, 142
on molecular phylogeny, *140*
not a single lineage, 140
sensing (other than echolocation), 134
*Microraptor*, 113, *114*, 115, 120, 121
migration distances, 7–8
minimum power speed, 54
mites, oribatid (*de novo* joint evolution), 80
models, testing of
extinct gliding lizards, 66
hypothetical ancestral insect, 82–85
silverfish falling, 87
molting
feather replacement and flightlesness in birds, 165
insect, 92
insect wings incapable of, 97
mayfly subimago, 92
Monarch butterfly migration distance, 7
Müller, Fritz, 77
murres, 13, 168
muscle tissue, 23
aerobic versus anaerobic, 24–25
fuels, 24–25
myoglobin in, 25
operating principles, 23
oxygen use, 24–25
muscles
antagonistic pairs, 23
constrain flapping, 24, 176
endotherms versus ectotherms, 121
evolution of insect wing and, 95
flapping, 111, 174
and hovering, 56
lactic acid and, 25
power for echolocation, 143
myoglobin, 25

naiad (immature aquatic insect), 78
natural selection, 21
    complex structures and, 23
    design in, 28
    for gliding, 88
    in gill theory, 77–79
    on living animals, 179
    on maneuverability and stability (flight), 27
    in paranotal lobe theory, 80
    sexual selection and, 81
    on solar collecting protowings, 84
Neoptera, 90, 95, 96, 97
neopterous wing articulation, 90, 96
Neornithes, 127
nervous systems
    active stabilizing by, 175
    controlling wing beat pattern, 25, 26
    directed aerial descent and, 49
    gliding and, 88
    of gliding animals, 61
    vertical take off reflex, 13
    wing-assisted incline running, evolution of, 120
nighthawks, 6, 15
Norberg, Ulla, 111
nymph (immature insect), 78, 98

Odonatoptera (including Odonata), 92, 93, 94
*Onychonycteris*, most primitive fossil bat, 133
"opposite birds". *See* Enantiornithes
oribatid mites (*de novo* joint evolution), 80
Ornithurae, *107*, *115*, 127
ostriches, loss of flight in, 166
Ostrom, John, 106, 109, 121
oxygen
    atmospheric, and giant insects, 94
    and bat respiratory systems, 145
    bird lungs obtaining, 123
    use by muscles, 24

Padian, Kevin, 149, 152, 154, *155*
Palaeodictyopterida, *91*, 92
Palaeoptera, 90, 92
paleopterous wing articulation, 90, 92. *See also* wing articulation
parachuting, 49, 80, 88

paranotal lobe theory (insects), 80–81, 85, 88
partridge chicks, and WAIR theory, 119
pelicans, 5, 15
penguins, 167–169
Pettigrew, John D., 137, 140
phylogenetic systematics, 16–19
phylogenetic trees. *See* phylogenies (phylogenetic trees)
phylogenies (phylogenetic trees), 16
    *Archaeopteryx* and birds, 106, *107*
    *Archaeopteryx-Dienonychus* relationship, 110
    arthropod limb structure and, 82
    assembling, 17
    of Avialae and theropods, by Godefroit team, *115*
    of bat families, molecular, 140
    bird-dinosaur relationships, 106, *107*, *115*
    computers and, 17
    convergence versus homology, 176
    examples, *18*
    genetic (DNA), 17
    insect-crustacean relationship shown by, 82
    insects on, *90*, 92
    Liaoning fossil beds, evidence from, 112
    of mammals based on anantomy, *132*
    of mammals, molecular, *133*
    Maniraptora and birds, 113
    megabats and microbats closely related, 140
    molecular, 17
    molecular, show microbats form two lineages, 140
    pouncing proavis theory and, 118–119
    pterosaurs and dinosaurs as sister groups, *152*
    pterosaurs, no molecular phylogenies for, 161
    reconstruction, 16–19
    separate pterosaurs from dinosaurs, 152, *153*
    shared derived characteristics and, 16
    splits megabats and microbats, 137

pigeon wing structure, 21
pilots, in air combat, 13
pinions (primary feathers). *See* feathers
pleural appendage theory, 77–79
plovers, 6, 15
pneumatic bones, 124, 159
pouncing proavis theory, 118–119
power
    for echolocation in bats, 143, 144
    endothermy and, 121
    for flapping, 24
    flight versus walking, 24
    and fuel economy in flight, 55
    and speed, 28, 53, 54
    U-shaped power curve of, 54
powered flight, 2, 50
    *Archaeopteryx*, 109
    bat evolution and, 143
    Enantiornithes, 127
    flapping, Big Four, 173
    gliding in the evolution of, 72
    thrust from flapping, 50–52
    two origins in bats, 137
powered flyers, "Big Four", 4–5
predation, maneuverability and, 175
predators, aerial, 14, 96
primary feathers, 30, *31*, 114, 126
primates, 131, 139
    on phylogeny, *132*, *133*
    vision, similarities with megabats, 139
*Protarchaeopteryx*, 112
prothoracic lobes of Palaeodictyopterida, *91*
protowings, 22, 111, 178. *See also* insect protowings
*Pteranodon ingens*, 9, 58
Ptéro-Dactyle, 148, 149
pterodactyls. *See* pterosaurs
*Pterodactylus*, 148, 149, *150*, 158
Pteropodidae (megabats), 137, 139, *140*, 141
pterosaur wing, *34*
    bones, 33
    stiffening fibers, 34
pterosaurs, *10*
    ancestry, 180
    arboreal flight evolution in, 154, 155, 174
    as archosaurs, 151
    bat-like reconstruction of, 149, *150*
    as bipedal runners, 154, *155*
    bird-like reconstruction of, *155*
    body mass estimates for, 159
    crests of, *160*
    cursorial evolution disputed, 155
    dinosaurs, relationship to, 151
    endothermy versus ectothermy, 159
    fingers of, 33
    fossil, *150*
        no transition, 149
        showing growth, 158
        trackways, 155, 156
    growth, size, and flight, 157–158
    *Lagosuchus* as ancestor, 152
    largest, *10*
    lungs of, 159
    origin of, 5
    phylogeny of, *152*, *153*, 161
    pneumatic bones of, 159
    "pterodactyl" as common name, 148
    Ptéro-Dactyle, first described, 149
    quadrupedal, 155
    *Scleromochlus*, link to dinosaurs, 151
    size, 57
    stability (flight), 27
    stiffening fibers, in wing membrane, 156
    swimming in, 161
    time of flight origin, 173
    walking in, 156, *157*
    webbed feet of, 161
    wing membrane of, 150, 154
    wing structure of, 33–35, 178
puffins, 13, 15
pygostyle, 125, 126

*Quetzalcoatlus*, *10*, 11, 159

Racey, Paul, 144
rails, loss of flight in, 167, 179
Rayner, Jeremy, 111
relative wind, 40
respiratory systems, 122, 145
Reynolds number, 45, 46, 158
Reynolds, Osborne, 46
*Rhyniognatha*, ancient insect fossil, 75

sandpipers, 6, 7, 15
*Scleromochlus*, 151, *153*

sexual selection, 81
shared derived characteristics, 16, 17
*Sharovipteryx*, 66, *67*
shoulder joint, birds' and ancestors', 123, 124, 126
shrews, *132*, *133*
silverfish, 75n, 76, 79, 81, *87*, 89
    falling behavior, 87
    primitively flightless, 100
size, 57
    and drag, 45
    growth, and pterosaur flight, 157–158
    hovering and, 12, 56
    insects versus vertebrates, 178
    largest flyers, 9
    limited by echolocation in bats, 144
    microbats versus megabats, *138*
    and protowing functions, insect, 84
    pterosaur body mass estimates, 159
    smallest flyers, 8
    and weight, 57
    wing structure and, 27
    and wings, effectiveness of, 85
smallest flyers, 8
snake, gliding, 68
soaring, 48, 58
Socha, John J., 68
social insects, winged and wingless individuals, 166
Soemmerring, Samuel Thomas von, 149
Solnhofen limestone, 103, 120, 149, 158
sonar. *See* echolocation
span, wing, 45
Speakman, John, 144
species in evolution, 16, 18
species naming, 11
speed
    for best fuel economy, 54
    flight advantage of, 3, 5
    flight, of large birds, 6
    glider, weight and, 47
    lift and, 43
    low aspect ratio wings and, 63
    maximum (flight), 6
    minimum power, 54
    and power, 28, 53
    trades off against distance, 28
springtails, 74, 75
stability (flight), 26, 27, 175

stability-configured airplanes, 26
stall, 41, *42*, 63
steering in dragonflies, 96
stiffness of insect wings, 35
stoneflies, 86–87
streamlining, 29, 42, 46
structure
    bat wings, *32–33*, 178
    bird wings, 30–*31*
    insect wings, 35–37, *36*
    pterosaur wings, 33–35, *34*
subimago (mayfly), 92
surface area, aerodynamic, 44, 56, 70
surface-skimming theory of insect flight origin, 86, 109
surface-to-volume ratio, 56, 71, 178
swans, 5
swifts, 14, 15
swimming in pterosaurs, 161

takeoff, vertical, 12–13
Taylor, Graham, 118
Teeling, Emma, 140
temperature-regulation. *See* thermoregulation
teratorn, 9
terminal velocity, 71
Teyler Museum and unrecognized *Archeopteryx* fossil, 110
thecodonts, 105, 151, *152*
thermoregulation, 117, 121
theropods, 106
    *Caudipteryx* and *Protarchaeopteryx*, 112
    and cursorial theory, 110
    endothermy in, 122
    evolutionary relationships of, *107*
    feathered, 113, 117
    phylogeny, including Avialae, *115*
    very small, 120, 121
Thomas, Adrian, 118
thrips, 8, 57
thrust, 50–52, 174
tracheal system (insects), 79, 94
tracks, fossil pterosaur, 155, 156
trees down theory. *See* arboreal theory
Troodontidae, *107*, *115*, 117
Tucker, Vance, 123
turbulent flow, 46
*Tyrannosaurus*, 117

unanswered questions
    bat and pterosaur ancestry, 180
    functions of protowings, 179
    timing of wing origins, 179
    wing precursors in insects, pterosaurs and bats, 179
uniramous limbs, 82
upstroke, *51*, 52, *53*
uropatagium (bat tailwing), 32
U-shaped power curve, 54

veins, insect wings, 35, *36*
*Velociraptor*, 106, 108
vertebrates, size, versus insects, 178
vertical takeoff, 12–13
viscosity, 46
vision
    in bats, 134
    requirement for flight, 175
    similarities in, megabats, colugos and primates, 139
*Volaticotherium*, 66
vultures, 45, 58

walking versus flight, 5, 174
wandering albatross, 8
warm-blooded animals. *See* endotherms
wasps, 15, 99
webbed feet in pterosaurs, 161
weight
    glider speed and, 47
    and hovering ability, 56
    and size, 57
    terminal velocity and, 71
wing articulation, 30
    *Confuciusornis*, 126
    dragonfly (interference), 96
    insect, 36, 89
        neopterous, 90
        origin, 180
        paleopterous, 90, 91, 92
wing extension, automatic, 31, 33
wing membrane
    bat, 32
    insect, 36
    pterosaur, *34*, 150, 154, 156
wing muscles and hovering, 54
wing pads (nymphs), 98
wing structure
    bat, 32–33, 178

    bird, *31*, 30–31
    insect, 35–37, *36*
    pterosaur, 33–35, *34*
    speed and, 28
wing-assisted incline running (WAIR), 119–120
winglets
    on model of hypothetical ancestral insect, *83*, 85
    of nymphs, 98
    Paleaeodictyopterida, 91
    on silverfish model, 87
wings
    area and lift on, 44
    articulation of, insect, 80
    aspect ratio of, 45
    basic attributes, 29, 30
    bat, 32–33
    bat, lift and thrust on, 33
    bat, membrane (patagium) of, 32
    bat, most primitive, 133
    bird, structure of, 30–31
    changing shape, 31
    cylinder acting as, 84, 85
    dragonfly, motions of, 96
    dragonfly, structure of, 21
    effectiveness and size, 85
    flapping forces on, 50, 52, *53*
    flapping movements of, *51*
    flapping, for thrust, 50, 174
    forewings and hindwings flapped together, 99
    gliding with, 47
    grasshopper, flapping pattern of, 95
    hinge interference, dragonfly, 96
    and historical constraints, 176
    homonomous, 95
    inability to molt, in insects, 93
    induced drag of, and tip effects, 44
    insect
        anatomy of, *36*
        four-winged plan, 94, 99
        legs did not become wings, 97
        neopterous hinge, 90
        paleopterous hinge, 90
        pleating, 35
    lift-to-drag ratio of, 44
    locked forwings and hindwings, 99
    modifications of bat front limb structure, 145

*Index* [209]

wings (*continued*)
    modified into halteres (flies), 100
    physics dictate form of, 176
    pigeon, structure of, 21
    pterosaur, *34*, *155*, 156
    scaling structure and weight of, 57
    size affects hovering, 56, 57
    of smallest flyers, 8
    stall resistance, 145
    stiffening fibers of, pterosaur, 34
    structural materials of, 21
    structure and scaling of, 27
    swimming with, 13

wingspan, largest, 8, 9
Wootton, Robin, 84
work of flapping, 24

*Xianglong*, 64

Yangochiroptera, *138*, *140*, 141
Yanoviak, Stephen, 48, 87, 88, 174
Yinochiroptera, 141
Yinpterochiroptera, *138*, *140*, 141

Zygentoma (silverfish), 76, 87